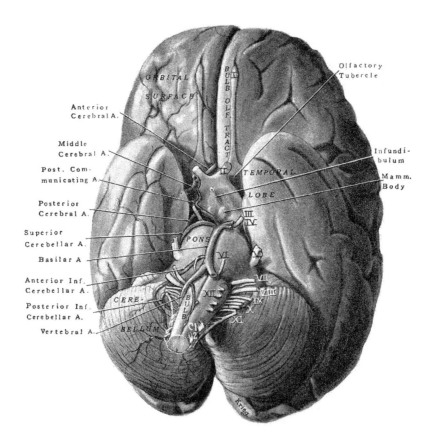

Anterior
Cerebral A.

Middle
Cerebral A.

Post. Com-
municating A.

Posterior
Cerebral A.

Superior
Cerebellar A.

Basilar A

Anterior Inf.
Cerebellar A.

Posterior Inf.
Cerebellar A.

Vertebral A.

ORBITAL
SURFACE

BULB
OLF.
TRACT

Olfactory
Tubercle

Infundi-
bulum

TEMPORAL

LOBE

Mamm.
Body

III
IV

PONS

V

VI

VII

CERE.

XII

BULB

VIII
IX

IX

BELLUM

Krieg

FRONTISPIECE: BASAL ASPECT OF THE HUMAN BRAIN

The arteries are represented on one side only. The cranial nerves are designated by
Roman numerals.

BRAIN
MECHANISMS
IN
DIACHROME

by

WENDELL J. S. KRIEG

B.S. IN MED., PH.D.

PROFESSOR OF ANATOMY,
FORMERLY PROFESSOR OF NEUROLOGY AND DIRECTOR
OF THE INSTITUTE OF NEUROLOGY,
NORTHWESTERN UNIVERSITY MEDICAL SCHOOL

Illustrated by the author

SECOND EDITION

BOX NINE   EVANSTON, ILL.

To my wife
ROBERTA
affectionately

20253

"To a man who knows nothing, mountains are mountains, waters are waters and trees are trees. When he has studied and knows a little, mountains are no longer mountains, waters are no longer waters, and trees no longer trees. But when he has thoroughly understood, mountains are once again mountains, waters are waters, and trees are trees."

PREFACE

NYONE who brings out a new book in neuroanatomy at the present day should justify its existence on the first page. A perusal of the reconstruction of the human brain in diachrome will supply the justification for this book. Only by a fortunate combination of circumstances was it possible to produce it, and to include it in a book of ordinary price.

The diachrome reconstruction will be found quite inclusive of the known structures and connections of the human brain, and the text is written chiefly around it. However, since all the material is discussed on the neural connections with which medical students and physicians are generally expected to be familiar, illustrations had to be designed for situations where the diachrome presentation was necessarily inadequate. The attempt was made to avoid increasing the number of illustrations, but to make each drawing as comprehensive as possible, with a resulting increase of compactness.

The textual approach is synthetic, systemic, functional, and the illustrations inclusive, graphic, three-dimensional. The approach assumes no special knowledge of anatomy nor of neuroanatomy, but the subject does develop rather rapidly. The needs and views of medical student, college student and physician can meet in the urge to build a clear picture of how the nervous system is put together and how it works.

Necessarily, in a small book with this approach, some topics must be abbreviated or omitted. This book does not illustrate or discuss actual appearances in microscopic sections (except for cortex), for these one must have recourse to the larger textbooks of neuroanatomy. Details of structure of neurons, nerves and neuroglia which are given fully in all textbooks of histology are not included; whereas blood vessels, meninges and many gross anatomical aspects are left to the textbooks and atlases of anatomy. All these considerations are sacrificed to the one object—a vivid three-dimensional visualization of the nervous system at work, as it progresses through animals and culminates in man—and this is badly enough needed.

This book attempts to give the cerebral cortex its proper significance and importance through a synthesis of our newer knowledge of its connections, the differentiation and architecture of its areas, and the results of localized stimulation at operation. Here the diachrome presentation is at its best, and few text illustrations were needed in this part of the book. The day is past when a sketchy treatment of the cerebral cortex is necessary or adequate. The cerebrum is the most important part of the brain from any standpoint—the image tube of the television receiver. Knowledge gained in the last decades, from a variety of approaches, can now be synthesized into a meaningful whole, although there is much, much more to be learned. It is chiefly in the field of the cerebral cortex that discoveries will be made within the lifetime of today's medical students.

This development has forced a shift in emphasis, and even in methods of study, in neuroanatomy. The time-honored kit of a dozen or two of Weigert sections, stopping at the superior colliculus, must now be supplemented by fiber dissections and sections demonstrating medullary fiber organization, diencephalic nuclei, cortical areas. The diachrome will be useful in guiding and interpreting the all-important fiber dissections.

The comparative approach should need no defense, although one has a feeling that with today's students it does. With the nerve components it is indispensable to real understanding. The presentation of the simplest generalized functioning brain mechanism, as shown in the salamander, can be a point of departure for the complexities of the human brain, can explain many of its features, and can furnish the "archi" stage of development. The inclusion of the rat reconstruction, matching the human diachrome, can supply a ready and graphic comparison of a simpler mammal with man, so that to a degree the human text can be used for study of the rat brain as well; the numbering and sequence of the structures in the reconstructions is the same. Yet those who do not wish to study the brains of lower animals can ignore the comparative chapter without any handicap in the study of subsequent chapters—the book has been planned with this alternative in mind.

The spinal cord is studied as a whole, as a simple model of the plan of the nervous system. After this, the cranial nerve mechanisms are studied, proceeding from the simpler motor neurons to the more complex vestibular system. Then the three main afferent systems are carried to the cortex. These are traced through their associational connections behind the central sulcus, then channelled through frontal areas and into the pyramidal system. The old motor system is given separate treatment, in ascending hierarchies. The cerebellar system is studied after all of these because it touches upon them all, and adds a quality to their effects.

The sequence of subjects, except for the inclusion of the comparative chapter, follows closely that of the author's much more complete text, "Functional Neuroanatomy". These two books can be efficiently used in conjunction.

More specifically, the book has been written to satisfy the needs of a variety of groups who study (or who should study) the connections and operation of the nervous system:

1. Primarily for the medical student as a supplementary or briefer textbook and set of illustrations in the neuroanatomy course; as a quick and graphic reference in his clinical courses, and in preparation for final or board examinations.

2. For the physician who is without laboratory material, and who wishes to supplant inadequate or outdated neuroanatomical preparation with brief text and graphic illustrations.

3. For dental students and other students of the special medical sciences whose time allotment to neuroanatomy does not permit study of a larger textbook.

4. For college courses, particularly for premedical students who wish to approach the human brain through the simpler brains of animals. Neuroanatomy, generally regarded as the most difficult subject in the medical curriculum, is more needful of a preparatory course in college than many of the "previews" commonly prosecuted.

5. For professors and students of psychology who want to know more about the brain than input and output, and are discontent with regarding it as "the black box."

6. And, hopefully, that educated persons generally can be led to take an interest in their most valuable and complex possession, the brain.

Chicago

22 June 1955

Wendell J. S. Krieg

PREFACE TO THE SECOND EDITION

It was not intended to bring out a second edition of this book so soon after its first appearance, but at this date the supply is exhausted. Here was the opportunity to correct an embarrassingly large number of wrong figure references and other ineptitudes. Once the printing forms are broken into one might as well revise the text and bring it down to date, and then add a few illustrations. When these were done it was apparent that a distinction had to be made between the two issues, though they can be used concurrently in class, without confusion.

The generous comments of the reviewers and the awarding of the Art Directors' Gold Medal of 1956 to the diachrome have contributed to the acceptance of the book, but probably this was partly because there is a place for a short and graphic textbook of neuroanatomy in medical schools, and for review by candidates, practitioners and specialists, as well as by psychologists and others who need a knowledge of neurology.

It would be ungrateful to preempt the space formerly occupied by acknowledge-ments and not to repeat that the opportunity offered by the Abbott International Company to create the diachrome for its overseas patrons, and that the extraordinary skill of Mr. Arnold Ryan, who traced and colored the author's reconstructions, were indispensable factors, without which this book could not have come into existence.

Whether another edition can ever be published when the present stock of dia-chromes is exhausted is problematic, as the unit cost of reprinting much less than a hundred thousand may be economically impractical.

For the information of those who are interested in the mechanical production of the diachrome: it is printed in rotogravure from a set of seven large cylinders, each printing a separate color component (red, yellow, two blues, black halftone, black line, and white), onto a continuous web-fed roll of cellulose acetate. All problems of register are compounded by the larger than usual number of colors, the "back-up" of the two sides of the folded sheet, and the through-and-through registry of the assembly. Confusion could have resulted were it not for especial care in positioning of the ele-ments in the original drawings, and the expertness of Mr. Ryan in the final *mise en scene*.

It has been an interesting and stimulating experience to write, illustrate, edit, design, and publish this book single-handed; and one which can be recommended to other professor-authors.

W. J. S. K.
1 October, 1957.

TABLE OF CONTENTS

Chapter One

BUILDING BLOCKS OF THE NERVOUS SYSTEM

ODIES OF all animals are formed of cells and their products, specialized in various ways to perform specific functions, and organized into tissues. Thus, epithelium is for protection, connective tissues for support, muscle for contraction, and nervous tissue for conduction of a stimulus. Although the organization of the brain reaches an unimaginable complexity, its functional part consists of only one type of cell unit, the neuron. All neurons have one or more thread-like extensions, enabling their property of conduction to span a distance which may reach a length of several feet. At some definite point the main body of the neuron is located; this is the vital center of the cell, and includes the cell nucleus. The extension is termed a nerve fiber. Generally, nerve fibers with similar connections and function are grouped together in bundles. When travelling among the non-neural structures of the body, they are called nerves, but when they are within the brain and spinal cord, they are tracts.

The nerve fiber is very delicate, so it must be supported by other tissues specialized for that purpose. Within a nerve (fig. 1: A), each nerve fiber is wrapped with fine but tough connective tissue fibers (endoneurium); bundles of reinforced nerve fibers are surrounded by a distinct outer wrapping (perineurium); and the bundles separated and bound together by further strengthening (epineurium). Nerves branch repeatedly, down to the ultimate nerve fibers, and these fibers may themselves branch, but nerve fibers never interconnect, nor form a true net. The branches of the nerve fiber from any nerve cell body are "private wires" through their entire course.

The terminations of nerve fibers may receive either stimulations from sense organs, or impulses from other neurons; or they may transmit impulses

[1]

to muscles or to other neurons. The cell prolongations (fig. 1: B) which receive impulses and conduct them toward the nerve cell are dendrites. Those which carry them away are axons. In nearly every situation, dendrites may be numerous and are essentially expansions of the cell body. For the tracing of impulses with which we shall be chiefly occupied, we can ignore the dendrites in the brain and consider axons as discharging on cell bodies.

The forms of endings of nerve fibers outside the brain and cord vary with the work they have to do (fig. 2). Each type of sense organ has its particular type of ending, and the muscular endings have a characteristic form. Sensory nerve endings are elaborated in various ways to permit maximum reception of the kind of stimulus they are to receive. Because there are so many sorts of sensory stimuli, each with its own conducting system to and within the brain, yet essentially only one sort of bodily reaction, muscular contraction, a greater part of our tract tracing will be on sensory or afferent paths than on motor or efferent paths.

The cell bodies, except those of the primary sensory neurons, are located within the brain and spinal cord, while these latter are a short way out on the nerve trunk, in little swellings called ganglia, generally one to a main nerve. Between the ganglion and the cord or brain is the sensory nerve root. Nerves to muscles do not possess ganglia; instead, the cell bodies are located in the cord or brain stem in dense accumulations called nuclei. In general, the cells of the intermediate or connecting neurons within the brain and cord are concentrated into nuclei, composed of like connections and functions, and connected by the tracts, already mentioned. Most of our study will be the location of these nuclei and the linking of them together by tracts. In two locations, but these are extensive, the multiplicity and special connections of the neurons are so great that they spread out to form extensive surfaces, these are the cortex of the cerebrum (figs. 41, 42) and the cortex of the cerebellum (fig. 43).

It is now clear that the neuron is the true structural and functional unit of the nervous system. The neurons arise, live, function and die as units. They begin life as round cells, out of which later grow their dendrites and their axons. If one of these extensions is cut, the distal part will die, but the cell body and other extensions may continue to live. Frequently, the cell body will show a change, when examined microscopically. In the brain and spinal cord, injured neurons never send out new nerve fibers. Functions destroyed thus never recover, although they may be partially replaced by intact systems with similar functions. In the nerves themselves, however, the fiber will grow into the tube left by the dead fiber, if the cut ends are close enough, and although the growth is only several millimeters a day, function can be reconstituted.

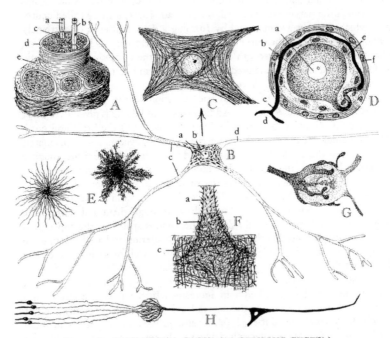

Figure 1. STRUCTURAL UNITS OF NERVOUS SYSTEM

A. Schematic drawing of a small nerve showing:
   a. nerve fiber
   b. neurilemma
   c. endoneurium
   d. perineurium
   e. epineurium
B. A motor nerve cell showing:
   a. nucleus
   b. Nissl granules
   c. dendrite
   d. axon
C. A nerve cell especially stained with silver to demonstrate neurofibrils
D. A spinal ganglion cell showing:
   a. nucleus
   b. finely dispersed Nissl granules
   c. unipolar process
   d. bifurcation into peripheral and central processes

e. satellite cells
f. capsular cells
E. Two neuroglia cells: on the left a fibrous astrocyte, on the right a protoplasmic astrocyte.
F. A pyramidal cell of the cerebral cortex indicating the wealth of connections it receives:
   a. end feet or boutons
   b. profusely branched axon terminals added
   c. diffuse axonal reticulum added
G. An especially intimate synapse, of the "private wire" type. The axon terminals grasp the secondary cell body.
H. A synapse in which the axon terminals from the olfactory cells on the left synapse with dendritic terminals of the mitral cell on the right.

The axons of most of the tracts and nerves are enclosed within a tube of fatty-protein nature, the myelin sheath, which serves as insulating material. Within the brain and cord there are none of the connective tissue sheaths with which the nerves are endowed, but instead, a special tissue called neuroglia, which achieves a similar end by forming a dense feltwork of extremely tenuous fibers among the nerve cells and their fibers (fig. 1: E). However, glia is unable to impart strength equal to that of the nerves, and if the brain and spinal cord were not enclosed within the skull and vertebral column, they would soon be destroyed. Severe injury to the cord by accident or tumor causes a permanent loss of control and sensation of the parts supplied by nerves joining at all lower levels. Additional strengthening is given by a thick membranous wrapping between the bony covering and the brain and cord. This is the dura mater, or dura. The actual limiting membrane, the pia mater, or pia, is thin and weak. They are separated by spiderweb-like tissue, tougher in certain locations, the arachnoid. Fluid is interposed between these coverings or meninges, which lubricates the surfaces subject to contact, and furnishes a cushion to absorb shock. This cerebrospinal fluid is secreted within the cavities or ventricles of the brain, flows through them and outside the brain, where it is absorbed, chiefly by the venous spaces at the top of the cranial cavity. By altering the relative rates of secretion and absorption, some compensation may be made for the tendency to swell or shrink, which the brain may show under altered physiological conditions. When there is hemorrhage into, or an infection of, the brain cavity, or an intracranial tumor, the pressure in this closed system increases. Intracranial pressure causes severe headache, distortion of the nerves leaving the skull, especially the one which moves the eyes outward (abducens, p. 43), causing a squint; and chokes off the optic nerve, causing changes in the retina of the eye and blurred vision; and if localized, can irritate or destroy brain mechanisms. When the circulation of the cerebrospinal fluid is obstructed, as by pressure on the narrow ventricular exits, there is no compensatory absorption, and the involved cavity swells causing pressure and resulting in symptoms, the nature of which will often give clues to the location of the disturbance. An increase of fluid inside or outside the ventricles is termed hydrocephalus, respectively, internal or external. In young people, chronic hydrocephalus causes permanent swelling of the skull itself, because the sutures between the bones have not yet joined.

Without a blood supply the brain could not function. There are two pairs of arteries to the brain (frontispiece). The vertebrals are branches of the subclavian in the neck, and enter through the foramen magnum, through which the spinal cord also passes. Under the brain stem the two vertebrals

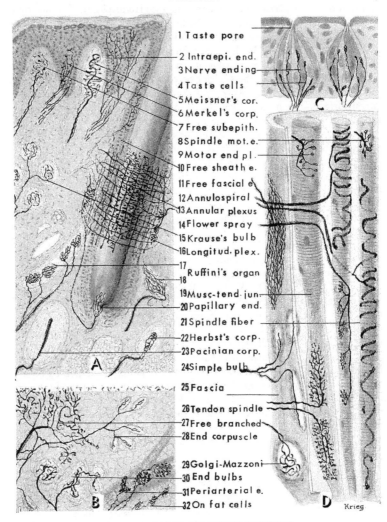

1 Taste pore
2 Intraepi. end.
3 Nerve ending
4 Taste cells
5 Meissner's cor.
6 Merkel's corp.
7 Free subepith.
8 Spindle mot. e.
9 Motor end pl.
10 Free sheath e.
11 Free fascial e.
12 Annulospiral
13 Annular plexus
14 Flower spray
15 Krause's bulb
16 Longitud. plex.
17
18 Ruffini's organ
19 Musc-tend. jun.
20 Papillary end.
21 Spindle fiber
22 Herbst's corp.
23 Pacinian corp.
24 Simple bulb
25 Fascia
26 Tendon spindle
27 Free branched
28 End corpuscle
29 Golgi-Mazzoni
30 End bulbs
31 Periarterial e.
32 On fat cells

Krieg

Figure 2. TYPES OF NERVE ENDINGS

A. In a strip through the skin
B. In visceral structures

C. In the taste buds
D. In muscle and fascia
   (assembled from various sources)

join to form the basilar artery in the midline (see frontispiece). This inverted Y sends off branches which supply the brain stem and cerebellum. The internal carotids are joined to each other and to the basilar, creating a continuous channel, the circle of Willis, which furnishes a considerable safety factor in arterial obstruction, by permitting alternative channels. The chief branches of the internal cartoid are the anterior cerebral, to the front and medial surfaces of the cerebral hemisphere, and the middle cerebral to the extensive lateral surface. The posterior cerebrals are the pair of terminal branches of the basilar, to the back part of the medial cerebral surface.

The venous drainage of the outside of the brain is through irregular and extensively communicating channels, but the blood is led away through venous sinuses, which are clefts in the dura mater. For the positions of these, and further numerous details of the arterial supply, text-books of gross anatomy should be consulted. There are no lymphatics in the brain.

Fracture of the skull may sever blood vessels, causing hemorrhage, usually by tearing the dura and its enclosed arteries. In persons with arteriosclerosis one of the minute arteries within the base of the brain may rupture, causing a hemorrhage within the brain tissue. Unfortunately, this usually afflicts the region where the important bundles of fibers controlling motor activity are travelling, producing a hemiplegia, or "stroke."

THE NEURON AT WORK. When adequately stimulated, a nerve fiber discharges a nerve impulse, a wave of negativity which proceeds along the neuron to its distal end, where it ordinarily is relayed by another neuron, or diffused among a number of neurons. A nerve fiber can only discharge, as a gun discharges when the trigger is squeezed. Strong stimuli may cause it to fire more frequently than weak stimuli, until its maximum rate is reached, but there is nothing specific about the nerve impulse, no coding to be interpreted at the opposite end. The specific nerve energies are entirely due to differences in the analyzing mechanisms. Large myelinated nerve fibers conduct more rapidly (100 meters per sec.) than small myelinated ones, and unmyelinated fibers conduct still more slowly (1 meter per sec.). Motor fibers are large, and pain fibers are small, but there is little correlation of size with function otherwise.

A synapse is a functional union formed between two neurons. Synapses vary in character. Axon terminals may end as tiny, bulbous expansions on nerve cell bodies or stalks of dendrites, and important motor nerve cells may be covered with these *"boutons"* (fig. 1: F). Sensory linkages may resemble a hand holding a ball (fig. 1: G). In other situations, axon and dendrite may make contact like two hands with fingers interlocked (fig. 1: H). In the cerebral cortex and elsewhere, the dendrites branch widely and make numerous contacts with the feltwork of axon terminals. When a nerve impulse reaches a synapse, there is a brief delay, of around one thousandth of a second. Brief as such intervals are, they can be detected with the cathode ray oscilloscope (p. 8). A chain of neurons operating together in response to one type of sensory stimulus, or directed toward functionally similar objectives, is a system. The study of functional neuroanatomy is simplified by grouping and tracing through the various systems. About sixteen are conventionally recognized.

In studying activity of the nervous system, we ordinarily consider that impulses pass up one or more sensory systems and down one or more motor systems. The simplest arrangement possible is a two-neuron reflex, where, for example, a pinprick sends an impulse through a sensory neuron, into the spinal cord, where it synapses with a series of motor cells, and sends down impulses which withdraw the limb. An intermediate neuron is interposed when an increased diffusion is desired, or a transfer to the opposite side or to another level. The number of neurons in an activity may be increased well beyond this, and when the cerebral cortex is involved, or all the secondary circuits considered, may be incalculable. Thus there are various levels of neural activity involving successively longer circuits, and one may speak of the spinal level, the mid-brain level, the cortical level, and so on.

METHODS OF INVESTIGATING THE BRAIN. ANATOMICAL. Without the application of special techniques the brain is little more than a soft, undifferentiated mass. When treated with hardening agents like formalin, and alcohol, it becomes possible to tease apart many of the fiber bundles and trace their course. For a study of the finer patterns of organization, tracing the fibers themselves and distinguishing nuclear accumulations, very thin stained sections mounted on glass slides must be studied under the microscope. Special staining methods have been developed for the nervous system, because the neuron resists the ordinary tissue stains. Each of the many nuclear accumulations has its cell pattern or portrait, and this may be brought out by certain blue dyes, collectively comprising the Nissl method, which stain the special flecks of nucleoprotein, Nissl substance, that are scattered through the cytoplasm (fig. 1: B). Nerve fibers may have an affinity for silver salts, and the silver may be precipitated by reduction similar to photographic development. This method reveals a network of tiny filaments within the cell body, the neurofibrils (fig. 1: C). This is the Cajal (pronounced "kahal") method. The most generally useful method, however, particularly for instruction, stains, not the neuron itself, but the myelin sheath; this is the Weigert method. Entire brains may be cut into series of sections by means of the microtome, a highly refined meat-slicing machine, and the nuclei and fiber tracts studied at leisure in permanent stained slides. The limiting factor is that the fiber growth is so luxuriant that individual fibers become lost in shifting from one section to another. Fortunately, an almost miraculous method has long been known, the Golgi (pronounced "Goaljee") method, which selects only scattered neurons and renders them in silhouette down to the ultimate ramifications of all their processes (see figs. 40, 41, 43). With all its usefulness, this method permits axons to be followed only a short way. To trace them farther, advantage is taken of the fact that a severed nerve fiber and its myelin sheath degenerate distally, so lesions of the desired structures are made experimentally. The Marchi method stains the degenerating myelin an intense black, and the Cajal method shows the degeneration products of the axon itself. Retrograde degeneration of the cell body after axon section can be used in some cases to determine the parent nucleus of a tract. The brains of embryos and lower animals yield a simpler picture and smaller sections than adult man, so have been much used in furthering progress toward the goal of understanding the human brain.

PHYSIOLOGICAL. Anatomical studies demonstrate mechanisms, but to be understood, these mechanisms must be observed in operation. The classic way of doing this is to stimulate a structure, usually electrically, or the counterpart, to remove it, and in either case, to study the effects. Thus, stimulation of the motor area of the cerebral cortex may lead to movements of the fingers, while operative removal of the same area will cause paralysis of the fingers. The course of the tract which effects a function may be discerned by making localized experimental injuries at various levels. The actual passage or arrival of the nerve impulses may be detected by a delicate galvanometer placed on the structure to be studied. The modern cathode ray oscilloscope is a refined galvanometer which visualizes the deflection of an inertialess beam of electrons, visualized on a fluorescent screen. The electroencephalograph (EEG) utilizes the electrical wave accompanying the passage of a nerve impulse to move an ink-writing lever instead of a cathode-ray. In the resting brain there is a characteristic rhythmic mass activity of the neurons, the alpha rhythm, which may be detected with the EEG. In epilepsy a different pattern is produced, which is most pronounced over the site of epileptogenic activity. By recording from various parts of the head, useful information may sometimes be obtained on the location and nature of disturbed brain activity; but the EEG is in no sense a "mind-reading machine," and the nature of the rhythmic activity is not understood.

Since presumably the brain of one animal of given species and size is identical with another, three-dimensional coordinate measurements locating any structure in the brain, taken from a series of slices of a sacrificed standard animal brain may be applied to a living animal. The stereotaxic machine is a mechanism for holding the head of an animal in a standard position, and for directing an electrode to a predetermined spot, with minimal destruction of overlying brain tissue. It has done much to integrate neuroanatomy and neurophysiology.

PSYCHOLOGICAL. Neurophysiological methods usually involve cutting of tissue, while psychological procedures generally imply an intact organism, although the boundary is not sharp, and the two methods may be combined. Psychological experimentation is characteristically applied to the analysis of sensations (input), and the study of the integrated reaction to stimuli or of brain activity (output). A knowledge of the mechanisms of the brain is badly needed by modern experimental psychologists; in fact, one of the principal purposes of this book is to supply to students of psychology an adequate amount of such knowledge in proper proportion of subject-matter. Psychological experimentation may be performed on animals or man. Sensory perception is analyzed subjectively or instrumentally, production and comprehension of speech is analyzed; learning and retention are studied, directly in man, indirectly in animals, by means of mazes, problem situations, and by instruments recording choices. Again, learnable drives and rewards may be utilized, or reflexes whose activity is conditioned by special stimuli, to obtain answers. Emotional reaction may be studied subjectively in man, or by observation of its bodily concomitants, as heart rate, skin resistance, in animals.

CLINICAL. The human nervous system would be but poorly understood were it not for the observations that have been made on diseased or injured patients by physicians who have correlated symptoms and clinical history with findings at operation or autopsy through the decades and interpreted their findings in publications. An animal cannot be commanded to make a specific movement, nor can it tell us what it senses when we make a specific stimulation. Voluntary action, precise manual skills, interpretation, subjective states, speech and reading, in short, any of the higher functions, may be investigated solely on man.

Neuroanatomy, neurophysiology, psychology and clinical neurology all furnish threads which, when woven together, give a pattern that enables us to understand the brain.

Chapter Two

TRUNK AND LIMB MECHANISMS: THE SPINAL CORD

ERVES that supply the motor control to, and relay sensory stimuli from, the trunk and the limbs arise from the spinal cord. The spinal cord is a column about half an inch thick, extending from the brain to the small of the back, and composed of nerve cells and fibers. Although itself delicate, it has a tough, membranous covering and is firmly enclosed within the bony vertebral canal formed by a series of bony, ring-like segments, the vertebrae (fig. 4). The nerves begin as two continuous series of tiny rootlets along each side of the cord (figs. 4, 5, 6), one toward the front, motor; the other toward the back, sensory. Opposite the space between each two vertebrae the rootlets are gathered into one round bundle on each side, the spinal nerves (figs. 4, 27). The nerves to thorax and abdomen remain separate to their terminations at the midline, in front, but those for the limbs, much heavier than the others, plait with one another to form plexuses (figs. 4, 30), the brachial plexus for the upper limb, the lumbo-sacral plexus for the lower limb. The several main nerves for the limbs arise from the plexuses and branch in a standard manner, each supplying definite structures. In the trunk the relations of the individual nerves to the skin regions (dermatomes) and the muscle-segments (myotomes), which they supply, is a simple and direct one; whereas the dermatome plan for the limbs requires research to trace. At all levels they are clinically important, because if the spinal cord is injured, one can determine precisely where the injury is located from the results of testing for the level of loss of sensation. Motor testing is much less definite. There is no point in describing the plan here; it is fully illustrated in figures 3 and 4, and explained in the accompanying legends, which should be studied in detail. Actually, the cord does not extend the full length of the vertebral column, but stops at the level of the second lumbar vertebra. The roots to lower intervertebral intervals continue down the vertebral canal, forming a horse's tail, or cauda equina (which means the same), and dropping off by pairs, pass between their appropriate vertebrae.

Our effort to understand the internal structure of the spinal cord will be aided if we have a glimpse of its development. The entire central nervous system begins as a tube which develops while the embryo is first discernible

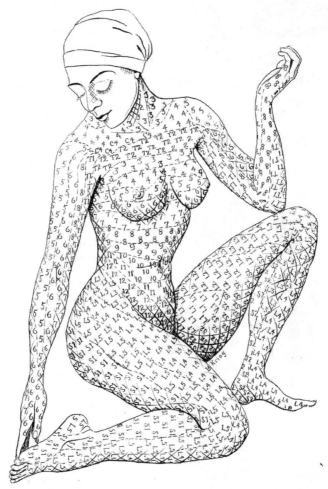

Figure 3. DISTRIBUTION OF DERMATOMES OF THE BODY

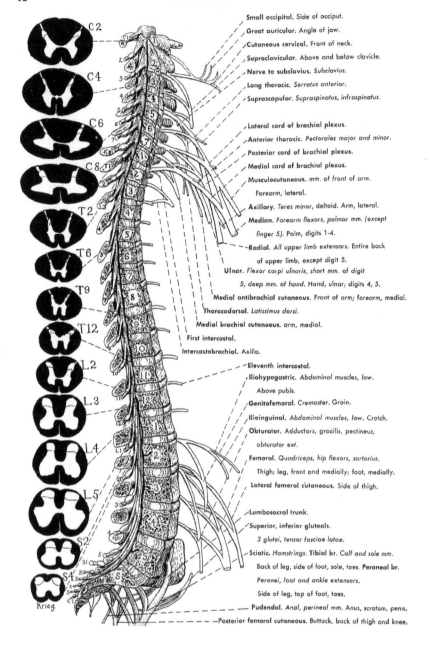

Small occipital. Side of occiput.

Great auricular. Angle of jaw.

Cutaneous cervical. Front of neck.

Supraclavicular. Above and below clavicle.

Nerve to subclavius. Subclavius.

Long thoracic. Serratus anterior.

Suprascapular. Supraspinatus, infraspinatus.

Lateral cord of brachial plexus.

Anterior thoracic. Pectorales major and minor.

Posterior cord of brachial plexus.

Medial cord of brachial plexus.

Musculocutaneous. mm. of front of arm.
Forearm, lateral.

Axillary. Teres minor, deltoid. Arm, lateral.

Median. Forearm flexors, palmar mm. (except
finger 5). Palm, digits 1-4.

Radial. All upper limb extensors. Entire back
of upper limb, except digit 5.

Ulnar. Flexor carpi ulnaris, short mm. of digit
5, deep mm. of hand. Hand, ulnar; digits 4, 5.

Medial antibrachial cutaneous. Front of arm; forearm, medial.

Thoracodorsal. Latissimus dorsi.

Medial brachial cutaneous. arm, medial.

First intercostal.

Intercostobrachial. Axilla.

Eleventh intercostal.

Iliohypogastric. Abdominal muscles, low.
Above pubis.

Genitofemoral. Cremaster. Groin.

Ilioinguinal. Abdominal muscles, low. Crotch.

Obturator. Adductors, gracilis, pectineus,
obturator ext.

Femoral. Quadriceps, hip flexors, sartorius.
Thigh; leg, front and medially; foot, medially.

Lateral femoral cutaneous. Side of thigh.

Lumbosacral trunk.

Superior, inferior gluteals.
3 glutei, tensor fasciae latae.

Sciatic. Hamstrings. Tibial br. Calf and sole mm.
Back of leg, side of foot, sole, toes. Peroneal br.
Peronei, foot and ankle extensors.
Side of leg, top of foot, toes.

Pudendal. Anal, perineal mm. Anus, scrotum, penis.

Posterior femoral cutaneous. Buttock, back of thigh and knee.

as a thickened disc, from a groove whose walls join together at the top and sink below the ectoderm. From the globular cells of this hollow tube, nerve fibers begin to grow outward into the tissues, forming the primary motor neurons. Other nerve fibers, on reaching the surface of the tube, turn and pass up or down or across the neural tube; they are connecting neurons or secondary neurons. The sensory neurons have a different origin. When the neural tube closed, a little line of cells was left between it and the ectoderm, the neural crest. Now divided into clumps corresponding to the myotomes, each cell emits two fibers, one toward the skin-to-be, the other toward the spinal cord. Throughout life these primary sensory cell bodies remain outside the cord, attached to the spinal nerve, as the sensory ganglia (figs. 8, 27). The dual fibers join to make one T-shaped connection to the

←

Figure 4. SPINAL CORD IN VERTEBRAL COLUMN: ITS CONFIGURATION: AND DISTRIBUTION OF SPINAL NERVES

The right half of the vertebral column is removed, revealing the spinal cord within the vertebral canal, and the nerve roots which leave it. Observe the gradually increasing mass of the vertebral centra as the column is descended, to the level where the weight is shifted to the pelvis.

The vertebral column is conventionally divided into several groups of vertebrae. In the neck are the seven cervical vertebrae. Their bodies are small, but the vertebral canal is large. In the thorax the vertebrae articulate with the ribs, constituting the twelve thoracic vertebrae. The five lumbar vertebrae are massive and rest on the sacrum, a wedge-shaped bone formed by the fusion of five vertebral elements. Between the vertebrae are the intervertebral fibrocartilages, cushion-like plates, which sometimes expand into the canal, paralyzing all structures below that vertebral level.

At each intervertebral interval a pair of spinal nerves emerges, formed by the fusion of two continuous series of rootlets: the motor in front, the sensory behind. At each intervertebral foramen or opening is the sensory ganglion. From the drawing it is obvious that the nerves and the ganglia differ widely in size. Those to the limbs (C5-T1, L2-S1) are larger than those to neck (C2-4), thorax (T2-L1) and buttock (S2-S5).

As the leaders from the cross-sectional diagrams reveal, the lower cord levels differ widely from the corresponding vertebral levels: From the middle of the thoracic region the remainder of the cord is foreshortened, the lumbar segments rapidly telescope, and the cord terminates just below L1. This means that the nerve roots must traverse successively longer distances in the vertebral canal, forming the cauda equina. It is to be remembered that pressure injuries below L1 cause no damage to the cord itself and can often be relieved by decompression.

By mastery of the principles of configuration of the cord in the text, and some knowledge of nerve distribution, the features of the various cross-sections may be explained, and the levels identified. For comparison the cord is best subdivided by the plexuses, rather than by vertebral groups. Opposite the cervical plexus (C2-C4) the grey matter is small and the white matter very large. Corresponding to the brachial plexus (C5-T1), the grey and the white matter are large; this is the cervical enlargement. Through the thoracic region (T2-T12) the grey is small, because each segment supplies only a narrow intercostal zone, but the white matter is medium. Opposite the lumbar plexus (L1-L4) the grey matter, particularly the sensory column, is very large, mostly from compression of the lumbar segments, whereas the white matter is small. Opposite the roots of the massive sciatic nerve the grey is particularly large. In the pudendal region (S3-S5), with its small nerves, the grey and white are both small.

On the right side of the drawing the roots of the various nerves and the constitution of the plexuses are shown. In the side headings the distribution of each nerve is summarized. Motor supply is set in italics, sensory supply in ordinary Roman. Mastery of this compact summary is an adequate knowledge of spinal nerve distribution for the practice of medicine. Since each nerve takes origin from nearly its entire plexus, there is little value in attempting to distinguish fine points in their levels of origin. The distribution of the dermatomes (fig. 3) gives the real level localizing data.

cell, which becomes large and round (fig. 1: D). When the cord-directed fiber reaches the cord, it bifurcates, one branch running toward the brain, the other in the opposite direction. The bifurcation of the primary sensory neuron is a general principle in both cord and brain, but there are exceptions. These longitudinally-running sensory axons send off side branches, collaterals, to cell bodies of neurons within the cord. These secondary neurons in turn send out axons capable of relaying the sensory impulse. For spinal reflexes they may pass in various ways, but the sensory impulses to be relayed to the brain follow the general plan of entering a secondary neuron which immediately crosses the midline, continues to the opposite lateral surface, and runs toward the brain. Each of these classes of secondary neurons, reflex and relay, have several important types, as we shall see presently.

Space demands cause the tube of cells to become virtually a solid column, but a tiny central canal remains in the center of the cord throughout life. The accumulation of so many longitudinally-running fibers enables a cross section of the developing or mature cord to be divided into two main parts: the inner column of cells, the grey matter; and the outer column of fibers, the white matter (figs. 4, 7). At any level, very few fibers are crossing the midplane, so the two halves of the cord are almost completely separated, the cells forming columns on either side, connected only by a thin strand. The forward or anterior half of the column is composed of the primary motor cells (motor column, fig. 7), in the rear or posterior half are the secondary sensory cells (sensory column, fig. 7). The size and configuration of each division varies with the local demands made by each spinal nerve, hence the outline of the grey matter varies from level to level, and may be used for approximate identification of level of any given section. The white matter accumulates as the brain is approached: there are both ascending sensory fibers to the brain, and descending motor control fibers from the brain. Hence the amount of white matter helps to identify the cord level. Opposite brachial and lumbosacral plexuses the cord is enlarged because more cells are needed to supply the limbs, while the mid-thoracic segments are reduced by the slight demands made upon them. A study of figure 4 will enable one to identify the approximate level of any section of the spinal cord.

REFLEXES WITHIN THE SPINAL CORD. Entering sensory neurons may join directly with a motor cell, forming a two-neuron reflex, exemplified by the control of muscular contraction by sensory fibers within the muscle (fig. 5: a, b). Reflexes with intermediate fibers may be intersegmental, ipsilateral (same-sided), and ascending (fig. 5: c, d, e) or descending (fig. 5: g, h, j); again, they may be intersegmental heterolateral (other-

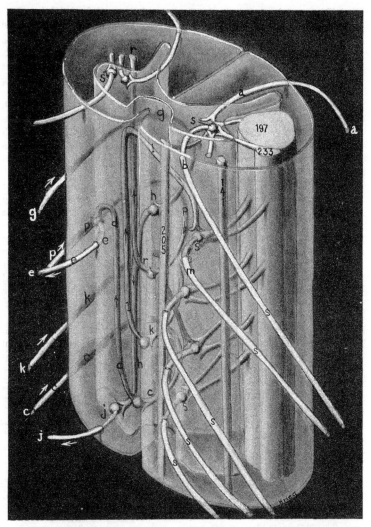

Figure 5. PHANTOM OF SECTION OF SPINAL CORD TRACING: ON LEFT,
SIMPLE REFLEXES, AND ON RIGHT, DESCENDING MOTOR CONNECTIONS

a, b. two-neuron reflex
c, d, e. ipsilateral ascending reflex
g, h, j. ipsilateral descending reflex
k, l, m. heterolateral intersegmental reflex
    p, r. part of a reflex connection running
        a long distance

s. motor neurons
141. tectospinal tract
197. lateral corticospinal tract
205. vestibulospinal tract
233. rubrospinal tract

sided) (fig. 5: k, l, m), and may run long distances (fig. 5: p, r). A great
deal of neural activity is mediated by connections intrinsic to the cord, and
running in the proprius bundles of the white matter, close to the grey and
deep to the long tracts (fig. 7). In fact, in cold-blooded vertebrates, posture,
progression and bodily reaction can proceed without aid of the brain, but
in mammals, particularly in man, the cord reflexes are under such a strong
influence from the brain that if the cord is cut, the severed part does not for
a time have enough potential to effect action, yet later the reflexes are much
brisker than normal, because uncontrolled. In this condition the reflexes
permit withdrawal of an injured limb and extension of the other as though
to take up the lost support; also the emptying of a full bladder or rectum
and other vital activities. In quadrupeds, standing and even walking movements are regained.

MOTOR CONTROL FROM THE BRAIN. The useful activity of
the muscles is controlled by the brain, largely through specific descending
neurons in named tracts. These will be detailed now, but their real nature
can hardly be clear until the brain connections are studied. The most important is the lateral or crossed corticospinal tract and its little brother, the
ventral or uncrossed corticospinal tract (p. 108). The former is located in
the dorsolateral white matter, and is very large (197: fig. 5; fig. 7), the latter
is ventromedially placed (fig. 7: V.Co-Sp.). They mediate the skilled and
willed movements of the extremities, particularly the fingers. Just below is the
rubrospinal tract (233: fig. 5; fig. 7: Ru. Sp.) from the red nucleus (p. 121).
It mediates basic control of posture.

The vestibulospinal tract (205: fig. 5; fig. 7) controls limb movements
and postures with relation to motion of the head. Its counterpart, the medial
longitudinal fasciculus (fig. 7, M.L.F.), establishes generalized torsion movements and postures. The tectospinal tract (141: fig. 5; fig. 7) mediates bodily

→

Figure 6. PHANTOM OF SECTION OF SPINAL CORD, TRACING COURSE OF
SENSORY NERVE CONNECTIONS

a. Primary pain and temperature conducting neurons
b. Lissauer's tract, carrying pain and temperature conducting primary neurons a
   short distance upward in cord
c. Secondary cell bodies for pain and temperature in head of dorsal or sensory
   column
d. Secondary pain and temperature fibers
   crossing to form:
e. Lateral spinothalamic tract, composed of
   upward-running secondary pain and
   temperature fibers
f. Bifurcation of primary sensory nerve
   fiber
g. Descending branch of primary fiber
h. Primary neurons of touch and pressure

j. Secondary cell body for touch and pressure in neck and body of secondary cell
   column
k. Secondary touch and pressure fibers
   crossing to form:
m. Ventral spinothalamic tract
n. Primary neurons of spinocerebellar system running considerably upward to
   synapse on secondary cells in sensory
   cell column whose axons turn laterally to
   form:
p. Dorsal spinocerebellar tract, and:
q. Ventral spinocerebellar tract
r. Primary fibers of discriminative system
   ascending in dorsal funiculus, synapsing
   only in brain

responses to visual stimuli, on the reflex level. In addition, there are reticulo-spinal fibers, more diffusely scattered, which are important in basic motor patterns, forming a necessary background for the willed movements. All these concentrate on the motor neuron (fig. 5: b, s) and by their algebraic summation determine its state of activity.

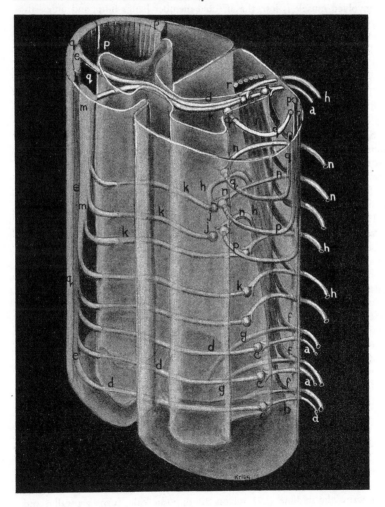

These tracts are grouped as upper motor neurons, the primary motor neurons that pass to the muscles being the lower motor neurons. There is a fundamental and clinically important difference between the results of the severance of the two. When the lower motor neuron is cut, all activity of the involved muscle is lost, the affected member hangs limp, and unless the nerve regenerates, the muscle will atrophy. When the upper motor neuron is severed, conscious control of the muscle is lost, and if all descending tracts are blocked, as in spinal cord compression, postural control is lost, too; but if the motor cell and nerve are alive and responsive, the muscle will not atrophy, and spinal reflexes become even heightened.

The short reflex fibers lie close to the grey, the longer ones further out, both together constituting the proprius (fig. 7). The long tracts, ascending and descending, lie outside on the ventrolateral portion.

SENSORY PATHWAYS TO THE BRAIN. (See fig. 6 for details.) Among the large myelinated sensory fibers that enter the dorsal root are scattered numerous small unmyelinated ones (fig. 6: a) which carry impulses incited by pain and temperature. Pain endings are widely distributed and of the free type (fig. 2: 2, 7). The lightly encapsulated tangles of Krause are considered cold receptors (fig. 2: 15); those of Ruffini, heat receptors (fig. 2: 17, 18). Their cell bodies are in the spinal ganglia and the axons enter a tract just external to the outer end of the column of sensory cells, known as Lissauer's tract (fig. 6: b; fig. 7: Liss.), but run upward only about one spinal segment before they turn inward to terminate in a special nucleus at the head of the sensory cell column, the gelatinosa (fig. 7: Gel.). The secondary cell bodies here, particularly small, may not give rise to long fibers directly, so may synapse again nearby; ostensibly, however, they send off a transversely-running long fiber which crosses the midline (fig. 6: d) and continues to the lateral periphery, then turns toward the brain and spans the entire distance to the thalamus, in the middle of the brain. The accumulation of similar fibers from all levels forms the lateral spinothalamic tract (fig. 6: e; fig. 7).

The other primary sensory fibers are coarser and bifurcate (f) on entering the cord, as stated. Both branches enter the posterior funiculus (fig. 7), medial to the sensory columns. The descending branch (g) relays data for reflexes at a lower level, but the ascending branches may run long distances. Those concerned with light or tickling touch (from intraepithelial endings, fig. 2: 6) and deep, painful pressure (from Pacinian corpuscles, fig. 2: 23) ascend for a variable distance, then turn into the body of the sensory cell column, where they synapse with a secondary cell (j). This cell sends its axon across the midline (k), but on the opposite side, turns ventrally, and on reaching the ventral surface turns toward the brain (m). Like the pre-

ceding, it continues quite to the thalamus, constituting the ventral spinothalamic tract (fig. 7). The sensations carried by the two spinothalamic tracts
are distress signals, carry some unpleasant affect, hence are grouped as the
nociceptive (pron. no-sigh-sep-tiv) system.

In contrast to this system, there are afferent tracts that carry impulses
of which we are never conscious. These come from endings in joints, tendons and fasciae (fig. 2: D), are large, heavily myelinated fibers, with cell
bodies in the spinal ganglia, which bifurcate and ascend, like the preceding,
and also terminate (n) in the sensory cell column. Those from the lower
limbs terminate in a conspicuous sub-column in the thoracic region, Clarke's

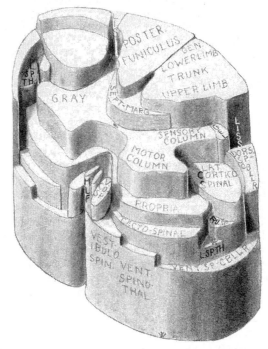

Figure 7. STEREOGRAM TO SHOW ARRANGEMENT OF SENSORY CELL
COLUMNS AND PRINCIPAL TRACTS OF SPINAL CORD

GEL.	Nucleus gelatinosus	Sept.-marg.	Septomarginal tract composed
LISS.	Lissauer's tract		of descending branches of pri-
L. SP. TH.	Lateral spinothalamic tract		mary fibers
M. L. F.	Medial longitudinal fasciculus	V. Co. Sp.	Ventral corticospinal tract
Ru. Sp.	Rubrospinal tract		

column. The secondary fibers (p) from the large cells there run laterally to the periphery, without crossing, then turn toward the brain. They may be followed into the cerebellum (diachrome 240: V. See p. 41), so are termed the dorsal spinocerebellar tract (fig. 7).

Other primary fibers, presumably entirely similar, synapse also in the sensory cell column, but not in Clarke's column, just in scattered cells. Their axons (q) also run laterally, but when they turn to run forward, take up a more ventral position. Some cross to the opposite side. Likewise, they end in the cerebellum but in the brain take a slightly different route (diachrome 241: V). This is the ventral spinocerebellar tract (fig. 7).

We are subjectively unaware of the impulses carried by the spinocere-bellar tracts. They furnish information on joint position, posture, and degree of strain to the cerebellum, which uses it to distribute muscular tonus through-out the body. Muscle, tendon and joint "sensations" are termed proprio-ceptive because they carry impulses proper to or originating within the body itself.

A third class of sensory fibers has an entirely different arrangement. These are the information-bearing fibers, without painful concomitants, which enable sensory syntheses, such as naming shapes, textures, denomina-tions of coins, together with some proprioceptive sensations which reach consciousness as muscle sensations, contributing to estimation of weight, size, shape. These are grouped as discriminative in opposition to nociceptive. Thus there are two types of proprioceptive fibers: unconscious, to the cerebellum; and conscious, to the cerebrum. The nociceptive fibers of the spinothalamic tracts, together with such fibers of this third class, whose endings lie near the surface, are grouped as exteroceptive. This third group, the discrimina-tive fibers, run to the brain as primary fibers in the dorsal funiculus (r). For this reason the dorsal funiculus increases progressively in size as the cord is ascended, and furnishes a good guide for identification of the level of a cross section of the cord (fig. 4). The ascending fibers from each segment become pushed medially by entering fibers from each higher segment. Precise bound-aries are maintained, so one may speak of a topological representation of the body segments which are lower than a given cross-section (fig. 7). In micro-scopic sections after degeneration from a crushed cord, one can estimate the site of the injury from examination of the funiculus; and the level of the section from a mastery of the information in figure 4. The further course of the discriminative system will be followed later. (p. 72)

In addition to sensations of exteroceptive and proprioceptive origin, there is a third main group, interoceptive, which deals with impulses from the bodily viscera. Most of these afferents are reflex-producing and seldom reach consciousness, and they are imperfectly understood. They run with the autonomic nerves (Chapter X), join the spinal nerves, and run upward with the nociceptive or spinothalamic system.

Because the various components of bodily sensation take separate pathways in the cord, localized lesions of the cord can lead to dissociation of sensation. The various components of sensation can be separately tested when examining a patient, as with pin-prick for pain, a feather for light touch, and a map can be made of the distribution of each type, permitting a diagnosis of the position and level of an injury. Light touch fibers cross at so variable a level that data are useless for diagnosis. In fact, if the cord is hemisected at any level, light touch is scarcely affected, because part of the fibers from any point are on one side as primary fibers, and on the other as secondary fibers. The student can now work out for himself the picture following hemisection of the cord at any level (Brown-Sequard's syndrome) (a syndrome is a group of symptoms indicative of one trouble).

INTEGRATION OF REFLEXES IN THE SPINAL CORD. It is now apparent that the activities of the spinal cord may be divided into two: the relaying of impulses to or from the brain, and the elaboration of intrinsic mechanisms. Before we can understand the former, we must learn the connections in the brain, but we are in a position now to study the latter. Most experimental work has been performed on the quadruped, and posture is easier to understand in a cat or dog than in a monkey or man. If the foot of a standing animal is pricked, the flexor muscles of the entire limb contract and the foot is withdrawn. This is the flexor reflex, an ipsilateral intersegmental reflex. At the same time, the weight must be shifted and a stronger contraction made by the extensor muscles of the opposite limb. This is the crossed extensor reflex, an intersegmental heterolateral reflex. (Fig. 7A, p. 22)

Quite different is extensor thrust, a contraction of the extensors of a limb induced by gentle contact of the sole with a solid surface. Basically, it is induced by the slight stretch applied to the tendons in the toes and is a manifestation of the widespread stretch reflex. In reduced form it is shown by the knee jerk: when the patellar tendon is stretched by tapping it, the extensor muscle attached to it contracts suddenly. If all four legs are reacting to the stretch reflex, we witness spinal standing. In general, whenever a postural muscle is placed under tension by the outer world, it contracts. In this way any threat to the *status quo* is opposed by muscular contraction, and normal posture is conserved. If a supporting limb is thrown far out of line, as in normal progression, when the body moves forward with the foot in place, a new reaction sets in: stepping. The foot is moved forward to a new position. Ordinarily this is developed in one leg of a pair at a time, while the other one develops an increased extensor thrust, brought on by the flexor pattern in the stepping leg. These are all the basic mechan-

isms necessary for posture and progression. Of course, balancing and coordination of movement must be supplied by the brain, so, to be useful, the
reflexes must be turned on or off by the mechanisms in the brain, and worked
in with the needs and desires of the animal. Quadrupeds whose spinal cord
is severed from the brain can stand and even progress imperfectly, but if
thrown off balance, they collapse and can never right themselves. The balancing and righting reflexes are in the brain. Spinal man, however, is a
pitiful object because the brain control is normally so dominant; yet the
mechanisms are present. Further on, we shall see just what contributions
the brain makes to posture, and shall come to realize that unconscious posture
is more complex than conscious movement.

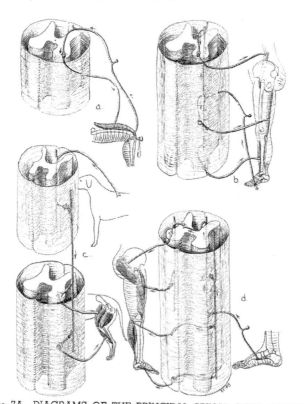

Figure 7A. DIAGRAMS OF THE PRINCIPAL SPINAL CORD REFLEXES
a. Knee-jerk   c. scratch reflex
b. flexor reflex   d. crossed extensor reflex
   (From Krieg: Functional Neuroanatomy. McGraw-Hill Book Co., New York. Second edition, 1953).

Chapter Three

THE EXPANDING BRAIN

RAINS of the various vertebrated animals differ as much in external appearance as do their bodies. Yet all are built on a common plan. In lower animals the plan is simple, to be sure, but structures and connections once formed tend to persist even to man, while new elaborations are formed from previously undifferentiated material. Hence we will be better able to understand the complexities of the human brain if we first survey the simpler brains and early stages in the development of the human brain.

A. THE CRANIAL NERVES OF THE BRAIN STEM: THE FISH

The simple arrangement of the spinal nerves does not continue into the brain; all structures of the head are supplied by a special set of nerves, the twelve pairs of cranial nerves. These are not arranged in the regular serial manner, but differ widely in nature and connections, requiring separate study, and forming a large part of the subject of neuroanatomy. Just as the spinal nerves were seen to be composed of sensory and motor components, so are the cranial nerves, but here the motor and sensory groups are each divided into three components. That is due to the diversity of origin of structures of the head and neck.

In development the great mass of skeletal or voluntary muscles of the trunk are derived from the somites, serially arranged segmental blocks of cells, one for each spinal nerve. The head has its somitic muscles, too, but they are confined to the muscles which move the eyeball, and those which move the tongue. They are innervated by somatic motor nerves. Although there are six eye muscles, they are derived from three somites, and they are supplied by three of the twelve cranial nerves (fig. 9: A, B): the third or oculomotor, fourth or trochlear, and sixth or abducent. The eye muscles are extremely conservative, there being no real change from fish to man. Several head somites are lost, so we drop to the twelfth or hypoglossal for the nerve supplying the tongue musculature. The hypoglossal has been forced, by the necessity of clearing the gill system, to arch backward to reach the floor of the mouth (fig. 9: A).

All the other muscles of the head are derived from the branchial or gill arches (fig. 9: C, D). Early in development of all vertebrates the unsegmented, non-somitic mesoderm at the site of the head forms into a series of bars, running from back to front, and these become pierced (in fishes; nearly

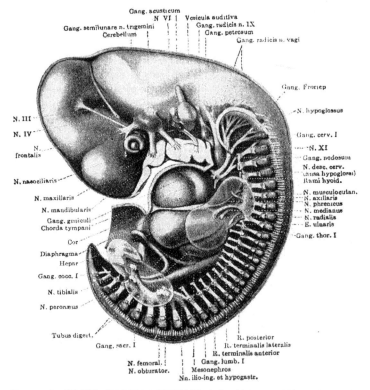

Figure 8. BRAIN, CRANIAL NERVES AND SPINAL NERVES OF A
15MM. HUMAN EMBRYO

(From Keibel and Mall: Manual of Human Embryology, J. Lippincott Co., Philadelphia, 1911)

so in the embryos of land animals) by slits between the bars. On the outer aspect is skin, on the inner aspect is mouth or pharynx lining; between is striated muscle, plus a rod of cartilage for muscular attachment and stiffening. Thus each branchial bar is a little sampler of tissues of varied origin, and a specific nerve component supplies each branchial arch (fig. 9: C), and each nerve contains one or an assortment of these components. The branchial arch muscles are supplied by the branchial motor component. Immediately forward of the gills proper is an outlet hole, the spiracle; and in front of this is the mouth, with separate cartilages for upper and lower jaws. The jaws are the first branchial arch. Its musculature becomes the jaw muscles in higher forms and is supplied by the branchial motor division of the fifth or trigeminal nerve. The second arch, associated with the spiracle, becomes the hyoid bone and musculature when the gills are no longer needed, in land forms. It is supplied by the seventh or facial nerve. In man, the greater part of the facial nerve supplies the muscles of expression on the face, formed also from the second arch. The third arch becomes the upper part of the pharynx and is supplied by the ninth or glossopharyngeal nerve. The fourth to seventh arches are supplied by one nerve, the tenth or vagus, which continues into the abdomen. In air breathing forms, these arches become pharynx and larynx. Thus, the conventional numbers of the cranial nerves do not correspond with the numbers of the branchial arches they supply: arches 1, 2, 3 and 4+ are supplied by cranial nerves V, VII, IX and X, respectively.

All the body is not skeletal muscle and skin: the viscera require a motor supply, both to the smooth muscle in their walls and to the glands associated with them. This is effected by a third motor component, the visceral motor component. It forms the autonomic nervous system, the subject of chapter 10, but its participation in the cranial nerves requires its introduction in chapter 4.

The nerves carrying sensations from the inner lining of the branchial arches form the visceral sensory component. The lining of the first arch, the mouth lining, however, is formed by a secondary ingrowth of the bodily covering, so does not carry a visceral sensory component. The VII (facial), IX (glossopharyngeal) and X (vagus), however, send their representation to linings of arches II and beyond. This distribution plan persists from fish to man. The sense of taste is a part of this system.

The outer or dermal coverings of all the branchial arches receive the somatic sensory or somesthetic component of nerves V, VII, IX, and X, as expected, and the lining of the mouth is also supplied by this component, through the trigeminal (fig. 9: E, F.). Actually all but the trigeminal are very small.

Three of the cranial nerves belong to the special sense organs (fig. 9: G, H). Each special sense organ is derived from a capsule. The first or olfactory goes to the olfactory epithelium of the nose, the second or optic connects with the retina of the eye, and the eighth or acoustic supplies the vestibular and cochlear sense organs of the ear. This leaves now only the eleventh or accessory, a sort of orphan among nerves. We shall regard it as branchial motor. The complete array of cranial nerves in man is shown in Table A, p. 59, analyzed by their components.

The brain accepts the theory of nerve components. The primitive brain stem is divisible into longitudinal strips, each contains the cells and fibers of one component (fig. 10). As in the spinal cord, motor is ventral, and sensory is dorsal. Through the middle part of the brain stem, however, the neural tube is apparently split open along the back and laid out, so that motor becomes medial, and sensory becomes lateral. In the motor section, the somatic motor nuclei are the most medial, the branchial motor are originally beside them (fig. 10), but in higher forms migrate to a more ventral position, so the visceral motor nuclei actually lie next to the somatic motor. Then follow the sensory components, with the visceral sensory medial, the somatic sensory next, and the special sensory most lateral (fig. 10). With a knowledge of how the nerve components are distributed among the cranial nerves, and their arrangement in the brain stem, the study of neuroanatomy becomes simplified.

B. THE PLAN OF THE FOREBRAIN:
THE SALAMANDER

The first or olfactory nerve, entering at the tip of the brain, is destined to have a far-reaching influence on the pattern of development of the cerebrum. Arising from neuroepithelial cells in the nasal mucous membrane, the unmyelinated olfactory nerve fibers synapse in the olfactory bulb (fig. 11: Ol. B.). From the olfactory bulb, secondary nerve fibers spread backward over most of the surface of the cerebral hemisphere directly behind. We shall divide them into three groups, the medial (fig. 12: med. olf.), intermediate and lateral (fig. 13: lat. olf.) olfactory striae passing respectively to its medial, basal, and lateral aspects. Strips are differentiated along the length of the cerebral hemisphere (fig. 11: CEREBRUM). On the medial surface are: the hippocampus, above, and the septum, below (fig. 11: HIPP., SEPT.). On the lateral surface are the pryiform cortex, above (fig. 13: PYR.), and the striatum, below (fig. 13: STR.). Basally is the

→

Figure 9. COMPARISON OF CRANIAL NERVE COMPONENTS
IN FISH AND MAN

The number of the cranial nerve represented in each component is indicated.

I. Olfactory	VI. Abducent
II. Optic	VII. Facial
III. Oculomotor	VIII. Acoustic
IV. Trochlear	IX. Glossopharyngeal
V. Trigeminal	X. Vagus
	XI. (Accessory)
	XII. Hypoglossal

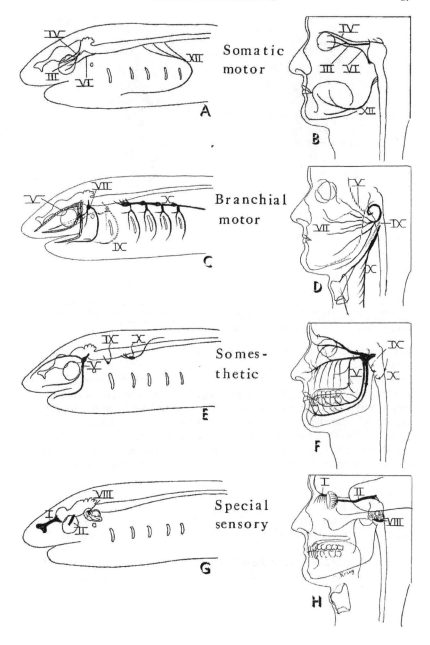

Somatic motor

Branchial motor

Somes-thetic

Special sensory

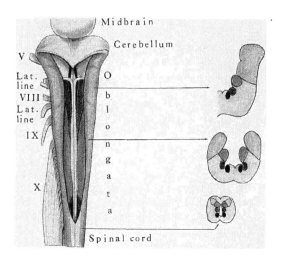

Figure 10. CELL COLUMNS IN BULB OF STURGEON

On the left a dorsal view, on the right, cross sections at three selected levels. Each cell column, representing a nerve component, is indicated by a different shade of gray, throughout the series; in order, from black: somatic motor, visceral motor, visceral sensory, somatic sensory. *(After Johnston)*

olfactory tubercle. The hippocampi in turn connect with the thalamic segment by two tracts: (1) with the habenula (fig. 11: HAB.) above, by the corticohabenular tract (fig. 12: co-hab.); and (2) with the hypothalamus (fig. 11: HTH.) below, by the fornix (fig. 12: fx.). The septum connects with the hypothalamus and brain stem by the medial forebrain bundle (fig. 12: m.f.b.). On the lateral surface the pyriform cortex and striatum connect with the thalamus and brain stem via the important lateral forebrain bundle (fig. 13: l.f.b.). Through this group of descending tracts, olfactory impulses can incite feeding activities, and exercise a certain amount of general control.

The other senses do not have a recognizable representation in the cerebrum, except for a somesthetic continuation by a primordial lemniscus which connects with the dorsal cortex (fig. 11: Do. Co.). This, the only non-olfactory cortex, is destined to expand enormously in higher forms as the neocortex.

Gustatory impulses enter the solitary tract (fig. 11: Sol.) through VII, IX and X nerves and ascend to the hypothalamus (fig. 12: 2° Gus.), where, meeting with descending olfactory impulses, they form a center for control of feeding and of visceral activities in general. The vestibular (fig. 13: VIII), impulses form direct reflex connections in the brain stem by the medial longitudinal fasciculus (fig. 12: m. l. f.), and with a cerebellum (fig. 11: CBLL), reduced here, though large in fish, which has an influence on distribution of muscle tonus. Many visceral and sensory impulses are satisfied by reflexes that go no higher than the bulb.

The optic system is a special case, because it is so large and because it enters the midbrain higher than all other nerves except the olfactory. The optic nerves cross completely below the hypothalamus and run dorsally along the outside of the thalamus as the optic tracts (fig. 13: II) to be distributed in the optic tectum (fig. 13), which forms a dome over the floor of the midbrain, or tegmentum (fig. 11: teg.). Athough unusually small in the salamander, in most other amphibia, the reptiles (fig. 14), and especially the birds (fig. 15), the tectum is so large as to form a major lobe of the brain, and has an elaborate structure comparable to that of the cerebral cortex of mammals. Obviously, vision is the most important sense in such animals, and the dominance of the visual center has attracted a strong representation of the somesthetic sense (fig. 12: lem.), the visual and somesthetic fibers ordinarily occupying individual layers, separated by cells. The tectum sends out the strong tectobulbar and tectospinal tract (fig. 13: te. bu., sp.), which runs down the side of the midbrain, crosses and distributes to the motor nuclei, exercising over them the highest integrated control. Tectal control of the eye muscles is an obvious connection; in addition there are tecto-thalamic, tecto-hypothalamic (te. hth.), and tecto-tegmental (te.-teg.) connections.

The connections of the cranial nerves are so similar to those of the fish that the previous description will suffice. However, since the gills are absent, the branchial nerves supply the structures developed from the former gill-arch material. See legend to figures 11-13 for additional details.

In the salamander there is virtually no differentiation or layering in the brain. The fibers form a layer on the outside of the tube, and the cells form a layer, locally thick or thin, on the inside, and their dendrites ramify in the fiber layer.

The brains of reptiles (fig. 14) and birds (fig. 15) are similar in plan to amphibians, with two major differences. The eyes are the dominant sense in most forms, and the other senses are reduced (except in birds, the vestibular). Hence, the optic tectum is enormous, tectal connections are increased, and tectal structure is complicated. As a result of olfactory suppression, the olfactory bulbs and the cerebral wall are much reduced, except for the striatal quadrant. The striatum increases in size until it forms a large bun in each

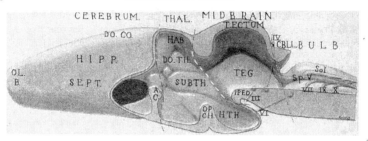

Figures 11 to 13. RECONSTRUCTION OF FORM AND CONNECTIONS OF SALAMANDER'S BRAIN SIGNIFICANT FOR UNDERSTANDING OF HUMAN BRAIN

Figures 11 and 12 are medial views; Figure 13 is a lateral view.

A. C.	Anterior commissure
AMYG.	Amygdala
CBLL.	Cerebellum
Co. hab.	Cortico-habenular tract
DO. CO.	Dorsal cortex
DO. TH.	Dorsal thalamus
fx.	Fornix
HAB.	Habenula
HIPP.	Hippocampus
h. p.	Habenulopeduncular tract
HTH	Hypothalamus
IPED	Interpeduncular nucleus
lat. olf.	Lateral olfactory tract
lem.	Medial lemniscus
l. f. b.	Lateral forebrain bundle
med. olf.	Medial olfactory tract
m. f. b.	Medial forebrain bundle
m. l. f.	Medial longitudinal fasciculus
OL. B.	Olfactory bulb
ol.-ped.	Olfactory peduncle
op. ch.	Optic chiasma
PYR.	Pyriform cortex
SEPT.	Septum
Sol.	Solitary tract
SP. V.	Spinal trigeminal tract
Sth.-bu.	Subthalamobulbar tract
St. Med.	Stria medullaris
STR.	Striatum
SUBTH.	Subthalamus
te. bu., sp.	Tectobulbar and tectospinal tract
TEG.	Tegmentum
te. hth.	Tectohypothalamic tract
te.-teg.	Tectotegmental tract
thal.	Thalamus
I	Olfactory nerve
III-XII	Cranial nerve nuclei and nerves
2° gus.	Secondary gustatory tract

A survey of the divisions and organizations of the salamander's brain is a good way of visualizing the most simplified plan of the brain. It is primitive and unspecialized; and it is the most thoroughly studied of any brain. Over 2,000 pages have been published concerning it, by C. J. Herrick during the last half century.

The salamander brain is a straight, elongated tube, bifid at its front end to form a pair of cerebral hemispheres (fig. 11: Cerebrum). In front they are capped by the olfactory bulbs (ol. b.), which receive the olfactory nerves. Between olfactory bulb and cerebral hemisphere proper is the collar-like counterpart of the olfactory tubercle of higher forms. The main part of the cerebral hemispheres is divisible into long sectors: hippocampus dorsomedially (hipp.), septum ventromedially (sept.), dorsal cortex dorsally and dorsolaterally (do. co.), while the ventrolateral sector is composed of striatum (str.) in front, amygdala (amyg.) in the middle, and pyriform cortex (pyr.) behind.

The remainder of the brain is a single

tube. The next segment is the thalamic segment (thal.), nearly split into two by the third ventricle. The dorsal thalamus (do. th.) is of restricted extent. It includes the habenula (hab.). The hypothalamus (hth.) below is longer than the thalamus. There are two conspicuous commissures below: the anterior commissure (a. c.), actually in front of it; and the optic chiasma (op. ch.) in the middle of its floor. The subthalamus (subth.) lies between dorsal thalamus and hypothalamus.

The next segment, the midbrain (fig. 11), contains the cerebral aqueduct. Its upper half is the tectum, its lower half is the tegmentum (teg.). Between the tectum and habenula is the posterior commissure. The medulla oblongata or bulb (bulb) completes the brain. Its rostral part is flattened and carries only a membranous roof except at its front boundary, where the cerebellum (cbll.) forms a diminutive ridge of nerve cells. In its lower part the two halves of the bulb fold upward and form a closed tube, which continues in that form throughout the spinal cord.

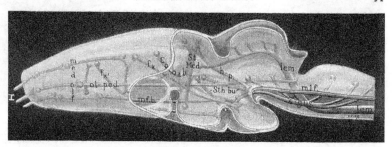

The motor cranial nerves are arranged on the same plan and supply substantially the same structures as they do in man, for the amphibians have attained terrestrial life, and have lost the gill mechanism. The somatic nerves supply only the eye muscles (fig. 11: III, IV, VI); the hypoglossal nerve has not yet separated from the spinal nerves.

The nerves to the branchial arches V, VII, IX, X (fig. 11) have been converted to their new role of supplying the oral and pharyngeal regions. The branchial motor components of these nerves supply the striated muscles which move mandible, pharynx and larynx; the visceral motor components control the smooth muscles and glands of the viscera; the visceral sensory components innervate the general sensory and taste endings of tongue and pharynx; while the skin of head and neck is supplied by the somatic sensory component. The last division is interesting here because the contributions of V, VII, IX and X nerves are nearly equal, whereas in man X is negligible and VII and IX are even questionable. The cell bodies of the neurons are arranged in columns disposed in the order named: somatic motor most medially, and somatic sensory laterally. (Special sensory, represented by VIII, lies more laterally still.) The bifurcating primary sensory fibers form longitudinal tracts, chiefly of descending nature. The somatic sensory neurons form a tract corresponding with the spinal V tract of higher forms, except that it receives a large contribution from VII, IX and X. The visceral sensory neurons form the solitary tract (sol.). Each tract is accompanied by a column of secondary neurons, which receive primary endings and send axons across the median plane, which thereupon turn rostrally to the midbrain and thalamus. The ascending somatic fibers join the general lemniscus (Lem.) to be described. The ascending visceral fibers form the secondary gustatory tract (2° gus.), which ends in midbrain and hypothalamus.

The vestibular nerve (fig. 13: VIII) enters the upper part of the bulb at its lateral edge, and bifurcates, sending descending branches which are distributed to the undifferentiated dendritic feltwork (neuropil) all down the bulb, and ascending branches to the rudimentary cerebellum. The cochlear division, large in man, is hardly discernible, but is developing from the dorsal division of VIII. Lateral line nerves, prominent in fish and amphibian embryos, are dwindling in the adult. Together they furnish the stimulus for developing the flocculi of the cerebellum, at its lateral ends. The cerebellum is unusually small in amblystoma, due to their sluggishness, but it receives also the spinocerebellar fibers and sends efferents which exert a tonic influence on the motor regions of the bulb.

The ascending sensory fibers of the spinal cord form the general lemniscus (lem), which roughly corresponds to the spinothalamic system of higher forms in that it carries crossed secondary somesthetic fibers. Few of them ever reach the thalamus; rather, they are distributed to the tectum, along with the optic fibers.

The optic nerves reach the brain at the cerebrothalamic junction (fig. 13: II), decussate (cross) completely at the chiasma and run dorsally along the sides of the thalamus as the optic tract, sending a few collaterals to thalamus and hypothalamus, but ending preponderantly in the optic tectum or colliculus. With the terminations of the two main space information systems, the tectum is the great center for discrete motor control. It develops 8 layers, showing more differentiation than any other part of the brain. Its main efferent is the large tectobulbar-tectospinal tract (fig. 13: te. bu., sp.), which drops ventrally into the tegmentum, crosses partially and passes down the brain-stem and cord, ending on motor cells along the way. The tectum also sends out tecto-thalamic and hypothalamic (te. hth.) and tectotegmental (te. teg.) connections, but the optic system has no connections with the cerebrum.

The important olfactory nerve (I) enters at the rostral end of the brain. The information it carries has a vital influence on feeding activities and produces a wealth of connections in the cerebrum. All the olfactory nerve

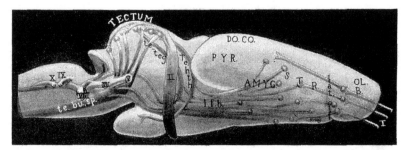

fibers synapse in the olfactory bulb. From here arise secondary fibers, running backward. As they flow back over the cerebrum, they encounter its various sectors, and these in turn give rise to important descending connections, which persist in the higher brain, often forced into indirect courses by unequal growth.

The hippocampus exchanges communications with the septum and sends down the cortico-habenular tract (co.-hab.), which combines with fibers from the preoptic region and septum to form the stria medullaris (st. med.), ending in the habenula. The habenula in turn sends the compact habenulo-peduncular tract (fig. 12: h.p.) to the interpeduncular nucleus (fig. 11: Iped.). This nucleus receives other descending olfactory elaborations, so is recognized as the final clearing-house of olfactory influences on somatic activity, operating through the motor control neurons of the tegmentum.

The septum and other ventromedial structures of the cerebrum send the medial forebrain bundle (fig. 12: m. f. b.) into the hypothalamus. Here olfactory and gustatory influences combine to institute a center for control of feeding and visceral activities.

The lateral olfactory tract (lat. olf.) reaches the lateral cerebral structures (fig. 13). The pyriform area remains olfactory in character, but the striatum is a somatic and olfactory correlating center of higher order and with motor connections. It gives rise to a large descending tract, the lateral forebrain bundle (l. f. b.), which ends in ventral structures all along the way: subthalamus, midbrain tegmentum and upper bulb. Some striatal connections end in the tectum.

The thalamus can represent little more than the median nuclei of the mammalian thalamus, as no sensory relay neurons to the cerebrum have developed. Its connections are numerous but unspecific. The subthalamus indicates its motor control nature by its crossed and uncrossed subthalamo-bulbar tract (sth.-bu.) to the tegmentum and bulb. The medial longitudinal fasciculus (fig. 12: m. l. f.) is the final link in generalized motor control. It carries effector impulses to the motor neurons of bulb and spinal cord, but does not contain the vestibular contribution so evident in higher forms.

For numerous further details and illustrations, see C. J. Herrick: The brain of the tiger salamander. (Univ. of Chicago Press, 1948).

cerebral hemisphere, while septum, hippocampus and general cortex are stretched and thinned almost to a membrane over the striatum, with the ventricular cavity a slit between. The striatum attracts somatic and optic connections and sends off a considerably augmented lateral forebrain bundle. The striatum becomes the highest center of motor control, apparently harboring the complex instinctual life of birds. Auditory connections have made an appearance; and the cerebellum is large and foliated in the birds. These latter increase in the mammals, but the optic tectum and striatum are never again so well developed.

Figure 14. OBLIQUE VIEW OF BRAIN OF A REPRESENTATIVE REPTILE,
THE ALLIGATOR

(From the Ziegler model)

C. CORTEX AND THALAMUS: THE RAT

The organization of the brain stem permits only reflex reactions; the expansion of the striatum achieves only stereotyped activity. Birds respond to cues, such as light and temperature, by complex behavior as migration and nesting. Mammals, whose activity is plastic, and who can learn from experience, have some obvious advantages, and the most teachable animal, man, dominates the animal world. The clue lies in the cerebral cortex, not the stereotyped olfactory cortex, but the somatic cortex, the expansion of the bit on the top of the salamander's brain that admitted somatic sensation.

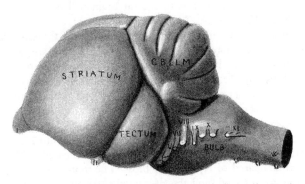

Figure 15. OBLIQUE VIEW OF BRAIN OF A BIRD, THE PIGEON
(From the Ziegler model)

The rat is a high enough mammal that we can see a definitive somatic cortex, separable into the main divisions which we find in more complex form in man; yet this animal is low enough to evade some of the complexities of the human brain. The poor vision of the albino rat reduces the size of the visual structures, equalizing the proportionate representations of the senses in the brain reconstruction. Considering also its availability, ease of rearing, activity and intelligence, it is the ideal form for study of the definitive plan of the brain, and it has been thoroughly worked out.

FIGURES 16 TO 23 FORM THE ATLAS OF RECONSTRUCTIONS OF THE RAT BRAIN, PLACED IN THE POCKET IN THE BACK COVER OF THIS BOOK

In the pocket inside the front cover of this book is the reconstruction of the human brain (explained on pp. 41 and 42), printed in diachrome; in the back pocket is a reconstruction of the rat brain, printed in halftone on paper. Both are of the right half of the brain, and corresponding structures are reconstructed in each similarly numbered sheet and given corresponding numerical designations. There are, however, twelve sheets of the human brain and eight of the rat brain, several having been combined. The rat brain reconstructions are to be used for the following description, which covers only general aspects of new features in the brain-plan, but the interested student will make comparisons with the rat as the human structures are being described in the remainder of this book and similar structures are identically numbered. Individual references to the rat brain will be given only rarely in chapters V to IX. (Described on p. 169ff.)

The central cranial nerve connections show almost the same pattern as in fish, strong as that statement may seem. The dendrites are now confined to the nuclei, the branchial motor nuclei have shifted to a more ventral position (87, 89, 91: fig. 16, 17); the hypoglossal nucleus is discrete (93) and a new system, the auditory, has developed from a part of the vestibular system. The details may be worked out from the human description (p. 60). Suffice it to say here that the connections follow the sequence: 207-214 in figures 20 and 21, and end in area 41 of the cortex in figure 23.

The striatum (126, 223-224) has expanded, compared with amphibia, but reduced compared to reptiles and birds. The motor controls established at that stage are retained, and the descending striatal tracts connect extensively in thalamus (136), subthalamus (162), and tegmentum (226).

The divisions of olfactory cortex in salamander are retained in a changed form. The hippocampus (230: fig. 22) becomes shaped and folded like a Parkerhouse roll, bent far back in the cerebrum. Its fornix (67: fig. 16, 17, 22) takes a curved course to reach the mammillary body (109: fig. 17).

The septum (66) and olfactory tubercle (T.O.) are conservative; the pyriform cortex is shifted backward (28: fig. 22). The old descending connections are there: the corticohabenular tract (joined with stria medullaris 107: fig. 17), the habenula (113) and habenulopeduncular tract (114) form one system. The medial forebrain bundle forms another.

The great change is that the entire dorsal and lateral surface of the cerebral cortex is of somatic connection and represents greatly expanded dorsal or somatic cortex of amphibia. Each of the three great sensory systems, bodily sensation, hearing, vision, packed with detailed information, has broken through to the cerebral cortex, has taken root, so to speak, spread, and

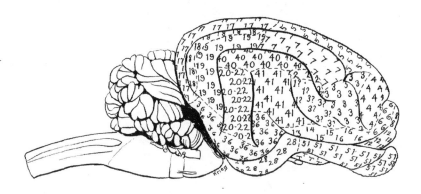

Figure 24. LATERAL VIEW OF BRAIN OF CAT, SHOWING
CORTICAL AREAS

budded off new connections. In turn the cortex sends numerous connections down to lower centers by way of the extensive internal capsule (figs. 20, 21) all along the way. No sensory fiber ascends directly to the cerebral cortex, all synapse in the thalamus. The somesthetic fibers synapse in the ventral thalamic nucleus (148: fig. 19), the auditory in the medial geniculate body (213: fig. 20), the optic in the lateral geniculate body (249: fig. 22). The combined bundle of the corticipetal (toward the cortex) and corticifugal (away from the cortex) forms a converging lamina, the internal capsule, which pierces the prominent striatum, dividing it (incompletely, in the rat) into a caudate nucleus medially (126: fig. 19) and a lenticular nucleus laterally (223, 224: fig. 21). As the internal capsule, now reduced in size from loss of the corticipetal fibers, runs along the lower surface of the brain stem, it is called the cerebral peduncle. Losing fibers along the way, it becomes the pyramidal tract in the bulb and crosses (196) to enter the cord as the corticospinal tract.

The active and varied life of mammals and the necessity for coordination with the cerebral cortex causes the cerebellum to develop large lateral lobes

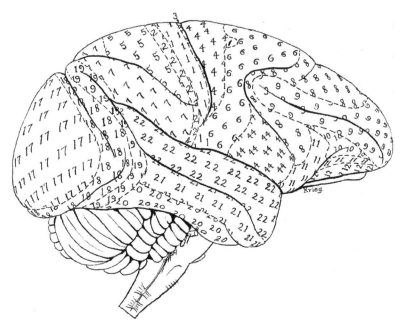

Figure 25. LATERAL VIEW OF BRAIN OF MONKEY SHOWING
CORTICAL AREAS

which are in neural connection with descending cortical fibers through the pons (188), which in turn sends the brachium pontis (244, 246: fig. 22) to the lateral lobes of the cerebellum (248). Comparative neurologists affix the neo- (new) to the parts which develop in mammals but are found only in token form in lower animals. Those structures present at the amphibian level are distinguished by the prefix archi- (ancient). Thus there are archi- and neostriatum, archi- and neocortex, archi- and neocerebellum, etc.

Among the widely divergent forms of mammals the chief differences in the non-cortical parts of the brain are of degrees of development of the

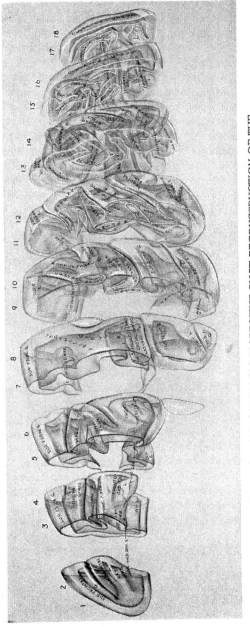

Figure 26. TRANSPARENT PERSPECTIVE SLICE RECONSTRUCTION OF THE
CEREBRAL CORTEX OF THE MONKEY

The outer and inner walls of the cerebral   is hidden in the sulci.
cortex are rendered as transparent sheets.   (*From Krieg: Connections of the frontal cortex of*
Gyri and sulci are labelled. Half the cortex   *the monkey. Charles C Thomas; Springfield, Ill, 1954*)

various systems. In the mole, for instance, the optic system is virtually absent; the duck-billed platypus has a large conducting system for mouth sensation; the whale has poor hearing, smell, and general sensation but a well developed vestibular system.

On the other hand, degree of cortical development differs widely. Animals whose bulk is large, and animals who are more intelligent, have a better developed cortex. In these types the cortex is thrown into convolutions which have a rather constant pattern in any species or group. In experimental work on the cerebral cortex, the convolutions are a guide to the site for stimulation or removal, so the pattern of its configuration becomes of importance in investigative work as the functional nature and connections vary with the locus. The extent of the folding is shown in the reconstruction of the monkey's cerebral cortex by slices (fig. 26), where the folded-in portion accounts for half the total. In man it is even more so. Ridges are termed gyri, valleys are sulci. A cortical region of similar function and connections is called a cortical area. These areas differ subtly under the microscope (fig. 42) and have been given standard numbers 1-52 to designate them. Their differentiation in the human brain will be a considerable part of future study, and a comparison can be made between rat (fig. 16, 23), cat (fig. 24), monkey (fig. 25) and man (diachrome I M, VI L). In the more lowly mammals the neocortex is exceeded by the olfactory cortex, located medially and laterally; in mammals of intermediate status the projection areas of the three senses occupy most of the neocortex, while in the series, monkey, apes and man, the associational areas, the frontal cortex and the temporal cortex increase steadily.

EMBRYOLOGY OF HUMAN BRAIN. In its development, the human brain sums up hastily and incompletely the history of the development of the brain in lower animals. Figure 8 represents the condition of the nervous system in a 15 mm. embryo. All the cranial and spinal nerves are present and may be studied in detail. The brain is divided into bulb, pons, midbrain, thalamus and cerebrum, the latter just beginning to form a vesicle on either side. Figure 27 is the brain and spinal cord of a human fetus of 135 mm. length. The cerebral hemispheres have become relatively enormous, have grown down over the thalamus and fused with it; and cover midbrain and part of the cerebellum, but do not fuse with them. As in phylogeny, the

$\rightarrow$

Figure 27. BRAIN AND SPINAL CORD OF HUMAN FETUS OF 135 MM.
(4 MONTHS)

Note the continuity of the insula, not yet covered by opercula, with the remainder of the cerebral cortex. The spinal cord has only begun to shorten to form the cauda equina; there is no secondary curvature; and the cervical and lumbar enlargements are clearly visible. The caliber of the nerves is correlated with the extent of the skin and muscles which they supply.

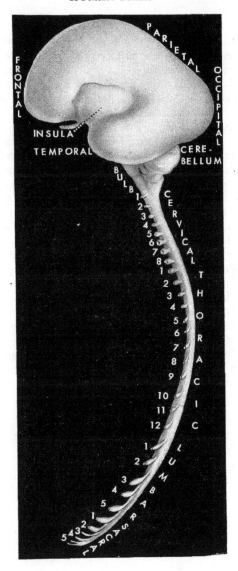

spinal and cranial nerve nuclei and primary connections develop first, then reflex connections, then other connections corresponding to the level of development of the amphibia, and finally the neocortex, neothalamus, and neocerebellum. Even within the sphere of the cerebrum the sensory projections to the cortex and the main motor pathway from it develop before the associational and other secondary connections.

A comparison of figures 21 and IV, L will show that a principal difference between the internal capsule of rat and man is that in the latter, a more extensive fan is formed. This is largely due to the development of the temporal lobe in man. The tendency of other structures of the human cerebrum to shift downward, then forward, is shown by comparing the position of the hippocampus in salamander (fig. 11), rat (230: fig. 22) and man (230, 231: fig. V M). Thus the fornix, at first a direct short connection (fig. 12), becomes quite arched (fig. 17, 22) and finally a complete circle (67: I L; 230: V M).

The brain is laid down as a tube, the neural tube, and it remains a tube throughout life. Unequal growth in the several segments causes it to form an elaborate ventricular system. Its form may be seen in the diachrome I L. In the bulb and pons it is rhomboidal and flattened as the fourth ventricle (75, 101), in the midbrain it is the narrow cerebral aqueduct (74), in the thalamic segment it becomes the slit-like third ventricle (71, 72), which communicates by the paired interventricular foramina (73) with the huge wishbone-shaped lateral ventricles, beginning with the anterior horn (100), continuing backwards through body (99), and posterior horn (98), then downward and forward as the inferior horn (97, connecting with 232: V).

USE OF THE DIACHROME RECONSTRUCTION

From this point onward, constant use will be made of the colored reconstruction of the human brain in diachrome in the pocket inside the front cover of this book. It should be before the reader throughout his study. It is futile to study the text without full use of the reconstruction. The subject is more easily understood if it is visualized and more easily remembered if a picture is recalled. The relations between the parts as shown, and their function as indicated by the color symbolism can channelize into all the student needs to know about brain connections for his work, and the reconstruction itself can then be utilized for a quick review of the subject. At this point the diachrome should be removed from the pocket and placed on the desk between you and the textbook. This position dodges the surface reflection.

The reconstruction is composed of six double sheets: twelve sides. It may be imagined as a box, the medial structures of sheet I forming the lid, and the lateral cerebral cortex in sheet VI being the box itself. Inside, sheets II-V represent a dissection, medial and lateral aspects of the same structures being represented on the two sides of any one sheet. Structures and connections are portrayed in the proper orientation and relations to one another on each sheet, and from one sheet to another without distortion. Structures are in proper mediolateral sequence from sheet to sheet, with certain definite exceptions.*

The sheets are numbered with Roman numerals I to VI, and they will be designated by the proper numeral in references. For the most part the same structures are shown on both sides of any one sheet, but where one or the other side of a sheet must be designated, the letter M will be suffixed to indicate the medial aspect, on the front of the sheet, and the letter L for the lateral aspect, on the back.

Structures are designated by common or Arabic numbers 1-287, and the usual reference to a structure will give its number followed by the sheet number, thus (258: V L) refers to the portrayal of the semilunar ganglion on the back of the fifth sheet. In general, the sequence of numbers begins with I M and ends with VI L, but the first 52 numbers designate the cortical areas of Brodmann, applying the numbers referring to the same areas as they were originally. With this exception, then

62– 95 are on I M
96–125 are on I L
126–151 are on II
152–181 are on III
182–222 are on IV
223–265 are on V
266–287 are on VI

In the text figures are given the same numbers as in the diachrome. The numerical index of the gate folded sheet with the diachrome booklet applies to the text figures as well. (Repeated on pages 158, 159)

---

*To preserve the unity of the cortical projection system, its radiations to the temporal lobe have been included on sheet IV and hence appear medial to the structures of sheet V, whereas they are actually lateral to them. One may imagine this contingent as having been pulled out to preserve the continuity of the projection system. Also, the cochlear root (206: IV) and cochlear nuclei (207, 208) seem medial to the restiform body (242:V), although in the brain they are really lateral. Again, this was planned to preserve the unity of a system. Likewise, the facial nerve (89, 90) should be lateral to the medial lemniscus (176).

Each functional unit is rendered in a different color. These colors are not entirely arbitrary, but applied according to a plan in which the hue is indicative of the function. Purely motor neurons are pure red, purely sensory systems are pure blue, but different shades are used for the several systems. In the cerebrum, associational areas are yellow and the degree of combination of associational nature with motor nature is indicated by the hue of the resulting orange. In the same way the degree of combination of associational with sensory nature is shown in the hue of green resulting from this blending. Olfactory structures are shown in violet and impure olfactory connections depicted in tones of purple and violet. The old motor system ranges from reddish brown to brownish-red. Cerebellar connections are in the brown group. Thalamic nuclei and projections carry the colors of the systems or cortical areas they serve. Using this scheme, there are not enough principal colors for each functional system. Several colors had to be used arbitrarily: the vestibular system is blue-green, the solitarius system is orange, and the hypothalamic nuclei are colored arbitrarily.

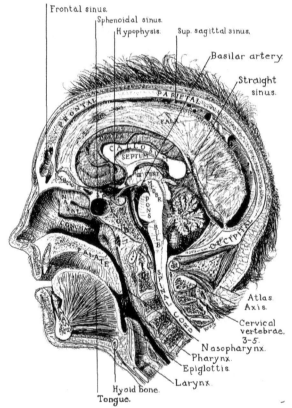

Figure 27A. MEDIAL VIEW OF BISECTED HEAD

(From Krieg: Functional Neuroanatomy. McGraw-Hill Book Co., New York. Second Edition, 1953)

Chapter Four

CRANIAL NERVE
MECHANISMS

CONTROL OF HEAD MOVEMENTS

OMPONENTS of the nerves to the muscles of the head and to the visceral muscles of the neck are arranged according to the general plan laid down in Chapter III: somatic, branchial and visceral motor components are represented.

MUSCLES DERIVED FROM SOMITES. All of the nuclei of the somatic motor system maintain their original position next to the ventricle and nearest of all to the midline. They are colored bright red in the diachrome. The family of reds is used for structures in the general motor or efferent category; the neurons closest to the muscle are brighter.

TONGUE MUSCLES. All of the muscles which have the suffix glossus in their name, that is, those which are inserted on or intrinsic to the tongue, are innervated from the hypoglossal nucleus (93: I; figs. 28, 32, 48, 49). At its lower end the nucleus is deep because the neural tube is closed and certain ventral structures have dropped out, but above, it is dorsal and on the floor of the fourth ventricle. The emergent fibers run directly ventrally, lateral to the medial lemniscus (176: III; figs. 32, 48), and between olive (255: V) and pyramidal tract (4: IV). Leaving the bulb on the lower surface (frontispiece, VI L: 265), the nerve passes through the skull at the hypogossal foramen, runs down the side of the pharynx and tongue and 'breaks up to supply its muscles (fig. 31: 14). In paralysis of one hypoglossal nerve, the tongue deviates to the paralyzed side when protruded.

EYE MUSCLES. Three nerves supply the six muscles which move the eyes. The abducent goes to the external rectus, the trochlear to the superior oblique, the oculomotor to the others, and to the levator palpebrae, which raises the eyelid. The abducens nucleus (88: I; figs. 28, 29, 32, 50) is located dorsomedially at the junction of the pons and bulb, within the genu of the facial nerve (90). Its long root passes directly ventrally and emerges at the pontobulbar junction near the midline (frontispiece, 88: fig. 32, 50).

The trochlear nucleus is located in the midbrain below the aqueduct

[43]

(86: I; figs. 29, 32). It is quite small and its fibers curve around the aqueduct (74. See I L). Reaching the dorsal surface of the midbrain, it crosses
the midline, and the tiny emergent nerve runs down the side of the midbrain
(frontispiece, figs. 29, 32).

The oculomotor nucleus (85: I; figs. 32, 51) is situated immediately in
front of the trochlear nucleus. It is larger, however, wedge-shaped, and so
close to the median line that the pair form one V-shaped unit. The coarse root
fascicles emerge between the peduncles, close to the median plane (frontispiece). All these eye muscle nerves gather behind the apex of the orbit (fig.
30) and pass through the superior orbital fissure and to the inner surfaces
of the muscles they supply.

The six extraocular muscles form three pairs. The medial and lateral
rectus turn the eye medially and laterally, respectively. The superior and
inferior rectus raise and lower the eyeball, but because they are obliquely
placed, the eye looks 20° medially when one contracts fully. The superior
and inferior obliques take care of the other of the three planes of space,
rotation of the eyeball on its own axis. They enable one to maintain a horizontal plane of reference when the head is skewed. In animals with laterally
directed eyes they are important; in man less so. Because they are inserted
behind the vertical axis of the eye, they also turn the eye to look laterally,
30° when fully contracted. The superior oblique turns the eye laterally and
downward while rotating the upper surface of the eye toward the midline;
the inferior oblique turns the eye laterally and upward, while rotating the
lower surface toward the midline. Superior rectus and inferior oblique, for
example, compensate for each other's deviation from the midplane. Paralysis
of ocular muscles can be detected by observing the compensating squint and
by asking the patient to look in various directions. The direction in which
one eye lags indicates the name of the paralyzed rectus muscle. Paralysis may
be either peripheral, in the course of the nerve, or central, in the nucleus or
internal root, a distinction with considerable difference in diagnosis. Differentiation between the two types depends on presence or absence of symptoms of damage to neighboring neural structures. The abducent nerve is
especially liable to irritation because of its long course between the pons and
skull-base. Paralysis or twitching of the external rectus is one of the classic
symptoms of intracranial pressure. Paralysis of an oblique muscle is difficult
to detect by direct inspection of the eye, but is indicated by the head being
cocked, to compensate. Disturbance of the conjugate movements of the
eyes, as forced lateral gaze, paralysis of lateral gaze, paralysis of upward gaze,
paralysis of normal convergence associated with viewing near objects, is a
sign of involvement of higher control pathways. The effector center for convergence, however, is believed to be the small nucleus of Perlia between the

oculomotor nuclei. It may be involved alone. Numerous studies have been made of the localization of the several muscles within the oculomotor nucleus. The results have been widely divergent, but there is clinical evidence that the order cephalo-caudally is: levator palpebrae, rectus superior, rectus medialis, oblique inferior, rectus inferior.

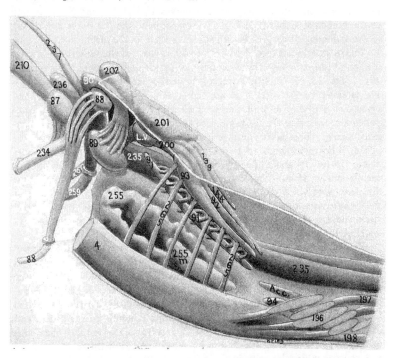

Figure 28. RECONSTRUCTION OF CERTAIN STRUCTURES OF RIGHT SIDE OF BULB AND PONS WITHIN A HALF-SHELL OF ITS OUTER WALL

4.	Pyramidal tract	200.	Vestibular nerve
87.	Masticator nucleus of trigeminal nerve	201.	Medial vestibular nucleus
88.	Abducent nucleus and root	202.	Superior vestibular nucleus
89.	Motor nucleus of facial nerve	210.	Lateral lemniscus
90.	Genu of facial nerve	234.	Root of trigeminal nerve
91.	Ambiguus nucleus	235.	Spinal trigeminal tract and nucleus
92.	Dorsal motor nucleus of vagus	236.	Main sensory trigeminal nucleus
93.	Hypoglossal nucleus	237.	Mesencephalic root of trigeminal
94.	Motor nuclei of highest spinal nerves	255.	Inferior olive
144.	Nucleus of solitary tract	255m.	Medial accessory olive
196.	Decussation of pyramids	259.	Facial nerve
197.	Lateral corticospinal tract	260.	Vestibular nerve
198.	Ventral corticospinal tract	265.	Emergent fibers of hypoglossal
199.	Descending vestibular root	L. V.	Lateral nucleus of vestibular
		Acc.	Nucleus of accessory nerve

MUSCLE DERIVED FROM BRANCHIAL ARCHES (see p. 24).
The cell column of the branchial motor component, originally just lateral
to the somatic motor column, becomes shifted to the ventrolateral part of
the reticular substance, but the facial and ambiguus units run back to their
original position before emerging. The nerves V, VII, IX, and X correspond
to the branchial arches 1, 2, 3, and 4, respectively, the last nerve continuing
to the abdomen. Rose pink is used for this system in the diachrome.

PHARYNX AND LARYNX. In the ventrolateral part of the reticular substance
is a long but extremely tenuous column of motor cells, the ambiguus nucleus
(91: I; figs. 28, 29, 32, 48). After a long or short recurrent course, the axons
run laterally to the surface of the bulb (263: V L) above the olive, and
emerge as a row of fine filaments which join the fibers of the other components to contribute to the vagus, and at the rostral end to the glossopharyn-
geus, nerves. The latter contingent passes only to one small muscle in the
palate. The vagus sends off two branches (fig. 31: 15, 23) to the muscles
of the larynx. Paralysis of this portion of the vagus, central or peripheral, is
usually unilateral and characterized by difficulty in swallowing and some
disturbance of voice production, which can be differentiated from other of
the many speech disturbances by examination with the laryngoscope, a small
mirror held at the back of the mouth. The slit-like opening becomes asym-
metrical, the vocal cleft deviating to the affected side.

FACIAL EXPRESSION. The numerous small muscles that move the features
are supplied by the facial or seventh cranial nerve. The facial nucleus is a
large ovoid collection of cells ventrolaterally placed, at the lower end of the
pons (89: I; figs. 28, 29, 32, 49, 50). Its efferent axons run at first dorsomedially, separately, to just behind the abducens nucleus, curve around it to
form the genu (90), then ventrally as a compact bundle to the caudal end
of the pons, where it emerges (259: V L; frontispiece), lateral to the abducent nerve. It then enters the internal acoustic meatus of the temporal bone,
and courses at first laterally in the bone (fig. 30) over the inner ear, then
turns sharply backwards, arching downward over the cavity of the middle
ear, and emerges from the temporal bone at the stylomastoid foramen. Within
the bone it sends off a nerve to the tiny stapedius muscle, which damps the
vibrations of the oval window (p. 60). Once on the face, it breaks up into
a radiating group of twigs which supply the orbicular and radial muscles of
eye, nose, mouth and ear, and the platysma — which are responsible for
movements of the features. At its exit, a deep branch supplies the stylohyoid
and posterior belly of the digastric, which aid in swallowing.

Spontaneous paralysis of the superficial branches of the facial nerve is
a common occurrence (Bell's palsy). Expression is lost on one side of the
face, the mouth droops, the nasiolabial furrow flattens, the eye cannot close

reflexly, and hence "waters," food gathers in the affected cheek. When peripheral in origin, the fibers are regenerated after some months; but one must be on the lookout for a central paralysis which is usually associated with signs of involvement of other parts of the facial nerve, or of brain stem structures.

CHEWING. The muscles used in mastication are supplied by the masticator division of the trigeminal or fifth cranial nerve. The masticator nucleus (87: I; figs. 28, 29, 32, 50) is a globular mass of motor cells at middle levels of the pons, dorsolaterally situated. Its root runs directly ventrolaterally to the surface of the pons far laterally (fig. 32). It runs under the main or sensory division of the trigeminal, bypassing the sensory semilunar ganglion, and enters the mandibular division, which, when it has passed through the foramen ovale, sends out a branch or two for each of the muscles of mastication (fig. 30). Paralysis of the masticator component causes the jaw to deviate to the same side when protrusion is attempted, but does not seriously hamper mastication. The uvula is deviated to the normal side from loss of tensor veli palatini.

ACCESSORY NERVE. The sternocleidomastoid muscle and upper part of the trapezius at the side of the neck are supplied by the (spinal) accessory nerve, which is probably of branchial arch origin. Its cell bodies are located farthest down in the brain stem (Acc.: fig. 28; 264: fig. 48) and even extend into the spinal cord. They form the lateral part of the ventral cell column, but the emergent fibers pass laterally instead of ventrally, gather into a collector trunk (264: V L; figs. 29, 48; frontispiece), which also carries some vagal fibers for a short way (vagal accessory). As the eleventh, or accessory, nerve it leaves through the jugular foramen with the ninth and tenth nerves, and enters the deep surface of the sternocleidomastoid muscle (fig. 31: 11). When paralyzed, the normal slope of the neck-shoulder junction becomes squared, the head cannot be turned to the opposite side, nor the shoulder raised normally.

VISCERAL MUSCLES AND GLANDS

The muscles whose innervation has been studied in this chapter are all of the striated or skeletal type. The walls of esophagus, stomach, intestines, bronchi, and arteries are composed of smooth or visceral muscle. The muscle forming the heart is of a type intermediate between the two. Smooth and cardiac muscles also receive a nerve supply, albeit a more diffuse one, not with a special nerve fiber and ending for each muscle fiber. Smooth muscle has a tendency to contract actively when stretched, and the viscera which contain it can operate without outside nerve control, but not in coordination with the requirements of the body generally. Hence a link with the central nervous system is needed. The nervous system which supplies the visceral structures, that is, all innervated structures but skeletal muscles and skin, is called the autonomic nervous system. Now, each organ (with exceptions) receives two types of autonomic nerve supply, one which stimulates activity, and one which inhibits activity; one nerve fiber cannot do both. These are the sympathetic and parasympathetic systems, *not* respectively, because the nature of the activity induced varies with the situation. The sympathetic system functions during stress or emergency; the parasympathetic during quiet, vegetative states. The two systems have quite different connections centrally, and it is because the greater part of the parasympathetic is closely associated with the nerves we have been studying that it is being introduced here. The remainder of the autonomic system will be discussed in chapter 10. The visceral motor component of the cranial nerves is parasympathetic. Its cell bodies, in the primitive position, are third in order from the medial plane, first and second being somatic motor and branchial motor, respectively. However, since the branchial motor cell bodies in the lower brain stem have migrated ventrally, the principal visceral motor nucleus, the dorsal motor nucleus of the vagus (92: I; figs. 28, 32, 48, 49) is alongside the hypoglossal nucleus. It is long and narrow, and coextensive with the hypoglossal nucleus. The axons are directed laterally and emerge as tiny rootlets along the middle of the lateral surface of the bulb (263: V; figs. 29, 32, 48), joining with fibers from the ambiguus nucleus and helping to form the vagus nerve. These fibers are distributed to the smooth muscle fibers and glands

→

Figure 29. PHANTOM OF BULB AND PONS IN DORSAL VIEW TO SHOW CERTAIN NUCLEI AND TRACTS

The structures reconstructed are imagined as enclosed within the hollow shell of the brain stem wall.

81. Inferior colliculus
86. Trochlear nucleus and nerve
87. Masticator nucleus of trigeminal nerve
88. Abducent nucleus
89. Motor nucleus of facial nerve
90. Genu of facial nerve
91. Ambiguus nucleus
92. Hypoglossal nucleus
93. Dorsal motor nucleus of vagus
95. Medial longitudinal fasciculus
144. Solitary tract and nucleus
165. Brachium conjunctivum
175. Medial lemniscus in pons
176. Medial lemniscus in midbrain
178. Nucleus gracilis
179. Nucleus cuneatus
199. Vestibular root
200. Spinal vestibular nucleus
201. Medial vestibular nucleus
202. Superior vestibular nucleus
207. Ventral cochlear nucleus
208. Dorsal cochlear nucleus
209. Trapezoid body
210. Lateral lemniscus
234. Trigeminal root
235. Spinal trigeminal tract
235n. Spinal trigeminal nucleus
236. Main sensory nucleus of trigeminal
237. Mesencephalic root of trigeminal
242. Restiform body
246. Brachium pontis
257. Trigeminal nerve
259. Facial nerve
260. Vestibular nerve
261. Cochlear nerve
262. Glossopharyngeal nerve
263. Vagus nerve
264. Accessory nerve
288. Vestibular fibers to m. l. f.
Int. Intermedius nerve
N. L. Nucleus of lateral lemniscus
S.A. Acoustic stria
S. L. Sulcus limitans
S. O. Superior olive
T. C. Attachment of chorioid roof of fourth ventricle

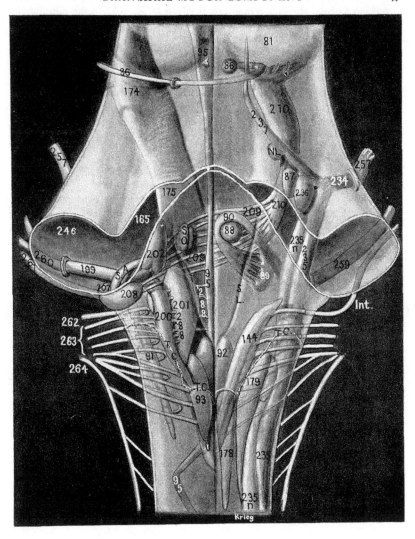

of the lungs and upper part of the abdomen, and to the heart. Autonomic nerves differ from other nerves in that there is always a second neuron interposed before the final termination is effected (fig. 46). In the distribution of the vagus this neuron is of microscopic extent. The cell bodies form microscopic ganglia scattered within the walls of the organ supplied. These cells and their short axons are termed postganglionic neurons. The neuron beginning in the dorsal motor nucleus and passing down the vagus nerve is the preganglionic neuron.

GLANDS OF THE HEAD. The visceral structures of the head are not reached by the vagus nerve so are supplied by whatever cranial nerve passes near them. The parotid salivary gland is supplied by the glossopharyngeal nerve. Just ahead of the dorsal motor nucleus of the vagus is a small nucleus near the floor of the fourth ventricle, and at the same transverse level as the other glossopharyngeal fibers, the inferior salivatory nucleus (I. S.: figs. 32, 49). The axons leave the glossopharyngeal nerve by its tympanic branch and reach the minute otic ganglion. Here are located the cell bodies of the postganglionic neurons. Two branchlets carry their axons to the parotid gland.

At the level of the facial nerve is another small nucleus, difficult to identify, the superior salivatory nucleus (fig. 32: S. S.). By a complex routing,* it supplies the remaining glands of the head.

MODIFICATION OF THE OCULAR IMAGE. Between the oculomotor nuclei, at the rostral end, is the small Edinger-Westphal nucleus. Its fibers pass out with the oculomotor nerve and reach the minute ciliary ganglion in the orbit (fig. 30). The postganglionic cell-bodies send the small ciliary nerves into the back of the eye (fig. 35). They run forward (1) to the ciliary muscle, which thickens the lens for close vision, and (2) to the sphincter pupillae, which reduces the amount of light entering the eye.

---

* One group leaves the brain stem in a separate bundle, the intermediate nerve, between auditory and facial nerves, joins the facial nerve in the internal auditory meatus, branches off via the chorda tympani, joins the lingual nerve, and synapses in the submaxillary ganglion. The postganglionic fibers pass to the submaxillary gland directly, and to the sublingual gland via the lingual nerve. The other group is routed through the intermedius and facial as far as the geniculate ganglion. Leaving via the greater superficial petrosal nerve, the axons, still preganglionic, join with the deep petrosal to form the nerve of the pterygoid canal and synapse in the sphenopalatine ganglion. Postganglionic axons run to the lacrimal gland via the zygomatic branch of the maxillary, to the nasal glands by short branches, and to the palatine glands by the palatine nerves.

VISCERAL SENSATION AND TASTE

Afferent fibers must be distinguished (p. 25) as to whether they come from branchial arch lining and viscera (visceral afferent); from skin, muscle and ligament (general somatic afferent); or from eye and ear (special somatic afferent). Each group is handled entirely separately in the brain.

Visceral sensation has a very different subjective quality from somatic sensation. The affective quality is more marked, and localization is poor. Distension of viscera causes severe pain, but cutting is painless. The endings of visceral sensory fibers are quite varied (fig. 2: B). They may be freely branched (fig. 2: 27), encapsulated corpuscles (28) or large bulbs (30). Pacinian corpuscles (23) are numerous in abdominal connective tissue. Afferent endings are found on arteries (31), which are viscera, even when they travel through somatic territory. Most visceral afferent impulses are never felt but are used as signals for unconscious reactions. Such are caused by (1) stretching of the bronchi during inspiration, reflexly arresting the inspiratory act; (2) distension of the aortic arch, introducing a reflex decrease in heart activity; (3) distension of the lower part of the internal carotid artery when the head is lowered, causing a reduction of blood supply to the brain; (4) registry of increase of blood carbon dioxide by the carotid body, increasing the respiratory rate. Other impulses reach consciousness, or at least cause obvious responses: (1) the presence of food in the pharynx induces swallowing; (2) gastric distress induces vomiting; (3) irritation of the larynx evokes coughing.

All these impulses ascend the vagus by fibers whose cell bodies are located in the nodose ganglion below the head (fig. 31: 9), or in the glossopharyngeal, in which case they enter its petrous ganglion. In either case, the sensory root reaches the side of the bulb, enters the solitary tract, and ends in its surrounding nucleus (144, orange: II; figs. 28, 29, 32, 48, 49). This tract and nucleus extend the entire length of the bulb. In the closed part of the neural tube below, they are dorsal to the central canal, being sensory; but in the open part they are lateral to all motor components.

Taste is a visceral sensation also, so is to be considered here. The taste bud (fig. 2: C) is a sensory ending, consisting of specialized epithelial cells filling a barrel-shaped cavity within the epithelium, among which nerve fibers wander. Most are on the vallate papillae at the back of the tongue; they are innervated by the glossopharyngeal. In adult man, only a few are on the anterior two-thirds of the tongue. Although fibers from here begin their brainward course in the lingual branch of the trigeminal (fig. 30), they leave it with the chorda tympani (fig. 30), which crosses behind the eardrum and runs to the bulbo-pontile junction as a part of the facial nerve. The primary cell bodies are located in the geniculate ganglion in the temporal bone.

There are a few taste buds in the epiglottis and pharynx which are supplied by the vagus. These three groups, enter the solitary tract, or, as some believe, by-pass it to enter an enlarged part of the solitary nucleus, medially, — the gustatory nucleus.

Taste is not the highly discriminative sense it is popularly regarded, for smell comes strongly into play in distinguishing flavors of food, as anyone can tell for himself when the olfactory endings are covered by mucus during a cold. Actually taste, in the narrower sense, is confined to the perception of sweet, sour, salt, and bitter. Some believe there is localization on the tongue for these several modalities.

Although the secondary cell bodies for visceral afferents are obvious enough in the solitary nucleus, their axons enter the reticular substance which forms the central core of the bulb and become lost as they are distributed to mediate the various bulbar reflexes. Thus, it is idle to look for any strong ascending sensory tract to the cerebral cortex. Taste, however, is now known to have a cortical localization; this is along the upper edge of the insula (266: VI M; fig. 52). The thalamic relay nucleus is perhaps the medial part of the arcuate (149: II L; figs. 37, 38, 39, 54).

COMPENSATION FOR HEAD MOVEMENTS AND POSITIONS: THE VESTIBULAR SYSTEM

The most ancient part of the anatomical complex we call the ear is the vestibule. In its most primitive form the vestibule is a fluid-filled capsule with nerve endings distributed around it to register the shift in position of contained granules as the head position is altered. This enables a primitive fish to keep righted in the dark. Very early three semicircular canals were evolved, one in each of the three conventional planes of space, and each with its sensory endings. Thus, while the animal turns, the flow of fluid, due to its inertia, is registered. Algebraic addition of the record of the three canals enables him to record and retrace his path, even in the dark ocean. The human vestibular organ (fig. 33) is almost identical with that of the fish. A membranous sac (membranous labyrinth) with three thin, hollow, semicircular canals is enclosed within a slightly larger cavity of bone (bony labyrinth), modelled to conform. At one end of each membranous canal is a swelling, the ampulla, with a ridge, the crista, which is covered with flagellated epithelium containing free nerve endings. When the head turns, the inerita of the contained fluid deflects the mass of hairs on the appropriate cristae. The membranous capsule itself is divided into two parts: one with its sensory epithelium or macula horizontal, the utricle; the other with its macula in the sagittal plane, pointing outward, the saccule. Any shift in head *position* is

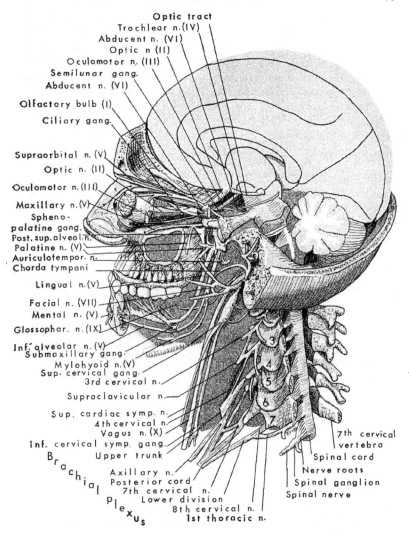

Optic tract
Trochlear n. (IV)
Abducent n. (VI)
Optic n (II)
Oculomotor n. (III)
Semilunar gang.
Abducent n. (VI)
Olfactory bulb (I)
Ciliary gang.
Supraorbital n. (V)
Optic n. (II)
Oculomotor n. (III)
Maxillary n. (V)
Spheno-
palatine gang.
Post. sup. alveol. n.
Palatine n. (V)
Auriculotempor. n.
Chorda tympani
Lingual n. (V)
Facial n. (VII)
Mental n. (V)
Glossophar. n. (IX)
Inf. alveolar n. (V)
Submaxillary gang.
Mylohyoid n. (V)
Sup. cervical gang.
3rd cervical n.
Supraclavicular n.
Sup. cardiac symp. n.
4th cervical n.
Vagus n. (X)
Inf. cervical symp. gang.
Upper trunk
Brachial
Axillary n.
Posterior cord
7th cervical n.
Lower division
8th cervical n.
1st thoracic n.
Plexus

7th cervical
vertebra
Spinal cord
Nerve roots
Spinal ganglion
Spinal nerve

Figure 30. DISSECTION OF CRANIAL AND CERVICAL SPINAL NERVES

recorded by the maculae. This is the way the head of a hen or rabbit is kept stationary and horizontal if its body is twisted about when held, and why the whites of the eyes show when one forcibly turns the head of a dog or cat. Man and all animals utilize vestibular impulses constantly to maintain a 0-0-0 position of the head and eyes. Let us study the neural mechanism.

For each of the five sensory epithelia there is a branch of the vestibular nerve (fig. 33: 1-6). They join at the vestibular ganglion nearby, in the fundus of the internal acoustic meatus. This and the adjacent spiral ganglion are the only primary sensory cell bodies which have retained their embryonic bipolarity. The vestibular root of the eighth cranial or acoustic nerve proceeds to the pontobulbar junction (frontispiece: VIII. 260: V, L; figs. 28, 29, 49), (the vestibular system is represented in bluish green in the diachrome) and runs dorsally through the pons to the lateral angle of the broad part of the fourth ventricle (199: IV L; figs. 29, 32, 49), where it bifurcates into ascending and descending roots, in the fashion of other primary sensory fibers. They discharge into a group of four secondary nuclei on the lateral half of the floor of the fourth ventricle. The ascending branch synapses with the secondary cells of the superior vestibular nucleus (202: IV; figs. 28, 29, 32, 49), and continues as primary fibers to the flocculus of the cerebellum (204: IV; fig. 49). Some secondary fibers of the superior vestibular nucleus

$\rightarrow$

Figure 31. COURSE AND BRANCHES OF VAGUS NERVE, AND OTHER NERVES OF HEAD AND THORAX

1. Ophthalmic and maxillary nerves, V
2. Trigeminal root, V
3. Mandibular, V
4. Internal carotid artery
5. Glossopharyngeal nerve, IX
6. Mastoid bone
7. Jugular foramen
8. Facial nerve, VII
9. Nodose ganglion, X
10. Hypoglossal nerve, cut
11. Accessory nerve, XI
12. First cervical nerve
13. Superior cervical ganglion, sympathetic
14. Hypoglossal, XII
15. Superior laryngeal nerve, X
16. Cervical sympathetic trunk
17. Pharyngeal plexus, X, sympathetic
18. Vagus trunk, X
19. Fourth cervical
20. Middle cervical ganglion, sympathetic
21. Thyroid gland
22. Middle cardiac, sympathetic
23. Recurrent laryngeal nerve, X
24. Ansa subclavia, sympathetic
25. Brachial plexus
26. Inferior cervical and stellate ganglion, sympathetic
27. Cut end of second rib
28. Inferior cardiac, sympathetic
29. Superior cardiac, X
30. Cardiac plexus
31. Recurrent laryngeal nerve, X
32. Inferior cardiac, X
33. Esophageal branches, X, sympathetic
34. Pulmonary plexus
35. Thoracic aorta
36. Esophagus
37. Accessory hemiazygos vein
38. Bronchial tree
39. Intercostal artery, seventh
40. Rami communicantes, T7
41. Cut end of seventh rib
42. Anterior esophageal plexus, X
43. Intercostal vein, artery, and nerve, T8
44. Hemiazygos vein
45. Splanchnic nerve
46. Sympathetic trunk
47. Left vagus
48. Sympathetic ganglion, T9
49. Sympathetic rami
50. Hepatic branches, X
51. Liver
52. Diaphragm
53. Hepatic plexus, X, sympathetic
54. Stomach
55. Gastric branches, X

(From Krieg: Functional Neuroanatomy. McGraw-Hill Book Co., New York. Second edition, 1953)

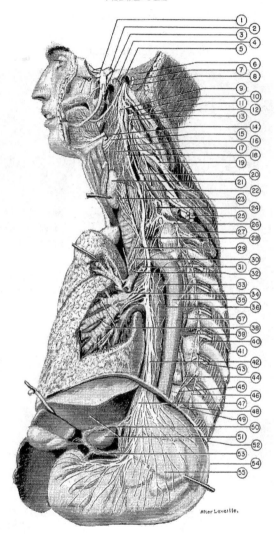

After Leveille.

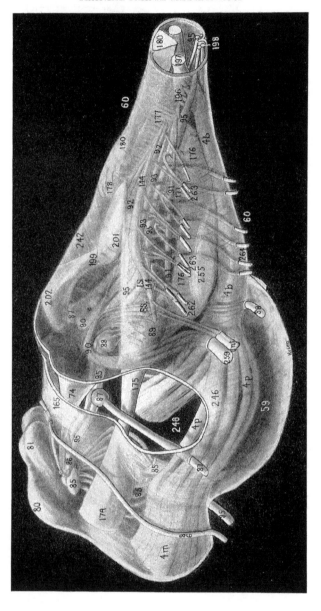

run to the fastigial nuclei of the cerebellum (203). The descending root (200: IV; figs. 28, 29, 49) forms separate fascicles, in the interstices of which are the cells of the spinal nucleus. Medial to it is the large prismatic medial vestibular nucleus (201: IV; figs. 28, 29, 32, 49), which receives a great share of the endings. The fourth nucleus is the lateral vestibular (Deiter's nucleus), large in mammals (rat brain: L.; fig. 21), but small in man, (fig. 49:205) situated at the bifurcation of the vestibular root. This secondary nucleus sends an ipsilateral tract down the ventral surface of the spinal cord (205: IV; figs. 5, 7, 48, 49), the vestibulospinal tract, which enables vestibular stimuli to increase or decrease the tone on the limb muscles.

The other three nuclei send fibers medially (288: fig. 29), to join the medial longitudinal fasciculus (often abbreviated M. L. F.), of the same and opposite sides, ascending and descending. This important tract (95: I; figs. 29, 32, 48-51), the most ancient of the brain stem, occupies the most medial and dorsal possible position. It extends the entire length of the brain stem, and continues in the cord, but it drops ventrally, because the neural tube has closed and the tracts which were ventral to it at higher levels have disappeared (fig. 7). It enables the vestibules to alter the tone of the axial musculature, but most specifically the eye muscles. Numerous endings pass to the oculomotor, trochlear and abducent nuclei.

Experimental stimulation of one vestibular nerve or MLF causes the eyes to be forcibly turned to the opposite side, and following this, the head, and the trunk. Reciprocally, cutting a vestibular nerve enables the opposite, intact vestibule to push eyes, head and trunk to its opposite side, *i.e.*, the affected side. Normally, the vestibules act together, the tonic effects of one

←

Figure 32. COMPREHENSIVE PHANTOM OF CERTAIN NUCLEI AND TRACTS WITHIN THE BRAIN STEM IN LATERO-DORSO-CAUDAL ASPECT

4m. Pyramidal fibers in cerebral peduncle	176. Medial lemniscus in midbrain
4p. Pyramidal fibers in pons	177. Internal arcuate fibers to medial lemniscus
4b. Pyramidal tract in bulb	
59. Pons	178. Nucleus gracilis
60. Bulb	180. Dorsal funiculus
74. Aqueduct	196. Decussation of pyramids
80. Superior colliculus	197. Lateral corticospinal tract
81. Inferior colliculus	198. Ventral corticospinal tract
85. Oculomotor nucleus and nerve	199. Vestibular root
86. Trochlear nerve	201. Medial vestibular nucleus
87. Masticator nucleus and root	202. Superior vestibular nucleus
88. Abducens nucleus and nerve	242. Restiform body
89. Facial nucleus	246. Brachium pontis
90. Genu of facial nerve	255. Inferior olive
91. Ambiguus nucleus	259. Facial nerve
92. Dorsal motor nucleus of vagus	262. Glossopharyngeal nerve
93. Hypoglossal nucleus	263. Vagus nerve
95. Medial longitudinal fasciculus	264. Hypoglossal nerve
144. Solitary tract and nucleus	Int. Intermedius nerve
165. Brachium conjunctivum	I. S. Inferior salivatory nucleus
174. Medial lemniscus in bulb	S. S. Superior salivatory nucleus
175. Medial lemniscus in pons	

balancing the tone from the other. When, for example, the head is turned to the right, the fluid currents flow to the left in the horizontal semicircular canal. Thus the right canal seems to push the eyes to the left. Similarly, movements in any plane stimulate some combination of semicircular canals and move the eyes in the opposite direction, with the purpose of keeping the eyes fixed where they were. Now, cristal, or canal, action is phasic, not tonic, and when the eyes are pushed as far as they can go, proprioceptive impulses from muscle stretch cause the eyes to be returned to the midline. The result is that with continued cristal stimulation the eyes oscillate back and forth in nystagmus. Nystagmus can be a symptom of vestibular irritation; the vestibular phase is slow, the recovery phase is rapid.

Similarly, the maculae of utricle and saccule induce the eyes to take up a new position when the head position is shifted, but, being tonic, it is not accompanied by nystagmus. The cristae are important during activity but the maculae are important for posture. In normal stance, macular stimulation is equal on both sides. When the head is tilted to one side, tonus of the limbs increases on that side, and relaxes on the other, as though to restore the threatened equilibrium. When the head is thrown back, the forelimbs increase tonus, the hindlimbs relax, as when a cat looks on a shelf; when the head is bent forward, tonus distribution is reversed, as when a cat looks in a mouse-hole. Vestibular reactions are more manifest in animals than in man. A rabbit is at the mercy of its vestibules; we are dominated by our eyes and suppress reflex drives by our persistent intellectualization. Nevertheless, the vestibule has a constant influence on posture. Anyone in the dark, or a blind person, cannot use the eyes for postural adjustment. The vestibules maintain the eyes in a horizontal and level plane, to a considerable extent the head also, and to some degree retain the perpendicularity of the body and distribute tonus to the limbs. However, as we shall learn later in more detail, once the eyes are centered, eye reflexes act on the neck, neck reflexes act on the trunk, and trunk on limbs in a connected sequence. The vestibules keep the eyes directed at a point while the body is turning, and in so doing establish data for orientation without visual cues. Although poorly developed in man, many animals have strong homing instincts or are able to get around well in darkness. They have no organs we do not, and they have large vestibules. Though the canals furnish sensory data for reflex compensation for centrifugal force while turning, by increasing the tonus of the outside limbs, the vestibule is not the "organ of balance"; that is more the function of the cerebellum. It is the organ for perception of the orientation and angular acceleration of the head. No one word covers this.

TABLE A: COMPONENTS OF HUMAN CRANIAL NERVES

No.	Nerve	Components*	Primary cell body	Course	Peripheral termination
I	Olfactory		Olfactory epithelium	Through roof of nasal cavity	Olfactory epithelium
II	Optic	SSS	Ganglionic layer of retina	Orbit → optic chiasma → optic tracts	Bipolar cells of retina → rods and cones
III	Oculomotor	SM	Oculomotor nuc.	Orbit	Rectus sup., inf., med.; obliquus inf., levator palpebrae mm.
		VM	Edinger-Westphal nuc.	Ciliary ganglion → ciliary nerves	Constrictor pupillae and ciliary mm. of eyeball
IV	Trochlear	SM	Trochlear nuc.	Orbit	Obliquus sup. m.
V	Trigeminal	BM	Masticator nuc.	With mandibular	Mm. of mastication
		GSS	Semilunar gang.	Ophthalmic, maxillary, mandibular brs.	Face, nose, mouth
		GSS	Mesencephalic nuc.	With mandibular and maxillary brs.	Proprioceptive to jaw muscles and tooth sockets
VI	Abducent	SM	Abducent nuc.	Under pons, into orbit	Rectus lateralis
VII	Facial	BM	Facial nuc.	Temporal bone, side of face	Mm. of expression, Hyoid elevators
		VM	Sup. salivatory nuc.	a. Greater superficial petrosal to spheno-palatine gang. b. Chorda tympani to submaxillary gang.	a. Glands of nose, palate, lacrimal gl. b. Submaxillary and sublingual gl.
		VS	Geniculate gang.	Chorda tympani	Anterior taste buds
VIII	Vestibular	SSS	Vestibular gang.	Int. acoustic meatus	Cristae of semicirc. canals, maculae of utr. & sac.
	Cochlear	SSS	Spiral gang.	Int. acoustic meatus	Organ of Corti
IX	Glosso-pharyngeal	BM	Ambiguus nuc.	Jugular foramen → side of pharynx	Sup. constrictor, Stylopharyngeus mm.
		VM	Inf. salivatory	Lesser superficial petrosal → otic gang. → auriculotemporal n.	Parotid gl.
		VS	Petrous gang.	Side of pharynx	Taste buds of vallate papillae
		GSS	Superior gang.	Side of pharynx	Auditory tube
X	Vagus	BM	Ambiguus nuc.	Recurrent and ext. br. of sup. laryngeal n.	Pharyngeal and laryngeal mm.
		VM	Dorsal motor nuc.	Along carotid a., esophagus, stomach	Viscera of thorax and abdomen
		VS	Nodose gang.	With motor	Viscera of thorax and abdomen
		GSS	Jugular gang.	Auricular br.	Pinna of ear
XI	Accessory	BM	Accessory nuc.	Side of neck	Sternomastoid
XII	Hypoglossal	SM	Hypoglossal nuc.	Side of tongue	Mm. of tongue

*BM, Branchial motor; GSS, General somatic sensory; SM, Somatic motor; SSS, Special somatic sensory; VM, Visceral motor; VS, Visceral sensory.

Chapter Five

RELAY OF SENSORY DATA

HE THREE great sensory systems — hearing, vision, bodily sensation — have the common trait that they project strongly to the cerebral cortex; in fact, they are the greatest factor in the formation of the neocortex, and dominate its organization, even in man. The greater part of their messages reach consciousness and are the chief data for man's knowledge. To this lively and brilliant trio we must subjoin the sense of smell—conservative and backward.

HEARING

Hearing is the newest of the senses. Reptiles show a beginning, in birds it is developed but limited, some mammals make much use of it for protection, and the development of articulate speech in man has elevated it to the utmost importance in communication and instruction. The sense organ for hearing, the cochlea, has been elaborated from a macular epithelium of the amphibia, and the accessory organs have been made of some old bones and muscles that were not needed when the "neo-jaw" developed.

ORGAN OF HEARING. Sound waves cause the ear drum (fig. 33) to vibrate. Stuck to the inner wall of the ear drum is a tiny bone, the malleus (fig. 33), expanding above the drum into a rounded head, which connects to the crown of a bone shaped like a molar tooth, the incus (fig. 33). One "root" or crus attaches to a very minute bone which looks for all the world like a stirrup, the stapes (fig. 33). These bones form a bent lever linkage oscillating with the ear drum. The foot plate of the stapes forcibly vibrates an inner diaphragm, much smaller than the outer, filling the oval window. These ossicles bridge the air-filled cavity of the middle ear, in the temporal bone and transmit the vibrations unchanged to the fluid-filled cavity of the inner ear. As the oval window moves in, a diaphragm below, the round window (fig. 33), moves out, for fluids are incompressible.

Attached onto the front of the vestibule is the cochlea (fig. 33), a tapered, tubular expansion of the inner ear that is coiled exactly like a snail. In the vestibule we found a thin membranous tube within the larger bony cavity. This condition obtains in the cochlea also, but the bony tube is divided throughout its extent into an upper level (scala vestibuli, figs. 33, 34) and a lower level (scala tympani, figs. 33, 34) by a bony plate extending halfway out from the axis (modiolus). The outer half of the partition is formed by the counterpart of the membranous tube of the vestibule, quite altered — the cochlear duct (figs. 33, 34). The floor of the cochlear duct is formed by a strong membrane, the basilar membrane (fig. 34), composed of connective tissue fibers stretched tautly from the edge of the bony flange to the outer wall. These are, paradoxically, short at the base of the cochlea and become gradually longer and looser toward the apex. They oscillate sympathetically with the sound vibrations, and there is good evidence that each tone has its place on the membrane. All that is required is a mechanism for transforming these local vibrations into a nerve impulse. This is effected by the organ of Corti (fig. 34: D for details). On the basilar membrane rests a strangely

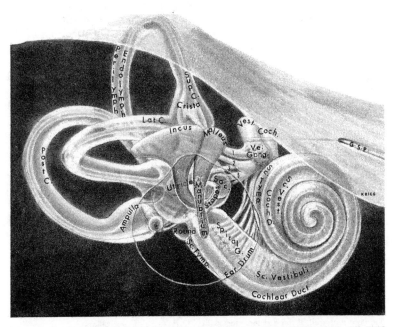

Figure 33. PRINCIPAL STRUCTURES OF MIDDLE AND INNER EAR OF THE RIGHT SIDE, IMAGINED AS TRANSPARENT, WITHIN PETROUS BONE, TOGETHER WITH DISTRIBUTION OF VESTIBULAR AND AUDITORY NERVES, IN VIEW DIRECTLY LATERAL TO HEAD

The labels are self-explanatory, except: G. S. P., greater superficial petrosal nerve.

modified columnar epithelium, covered with a firm, thin cuticle, the reticular membrane. This cuticle is firmly attached to the bony lip and kept under some tension by a row of inner pillars. The outer pillars are obliquely placed. Their base is over the middle of the fibers of the basilar membrane, their summit engages the top of the outer pillars, forming a strut for the reticular membrane. Thus, when the basilar membrane vibrates, the lift of the base of the outer pillars is transformed into a horizontal thrust, so the reticular membrane oscillates horizontally. Arching over the reticular membrane, and touching it is a gelatinous lamina, anchored at its base, the tectorial membrane (fig. 34: C). Into it are thrust a large number of minute filamentous structures arising from special hair cells attached at their crowns to the reticular membrane and seated upon the little seats formed by the phalangeal cells. Each cell of the latter type sends up a separate strut which expands into a flat plate, the whole arrangement being like a life guard's chair and its umbrella. These little umbrellas fit together by some of their sides to form the actual reticular membrane, but spaces are left for the tops of the hair cells. Thus a firm support is made for the delicate hair cells, and the horizontal oscillations of the hairs as they move relative to the tectorial membrane are transmitted to the hair cells. The hair cells are closely surrounded by a plexus of nerve fibers which are excited thereby.

Tracing back the nerve fibers, some echelon up the cochlear duct and between the rows of hair cells, phalangeal cells and pillars (fig. 34: NF), seeking the modiolus, and entering the bony lamina; others take a directly radial course. These are the dendrites of the cochlear nerve. Within the spiral central bony core of the cochlea the nerve fibers join their appropriate cell body, in the spiral ganglion (figs. 33, 34), composed of bipolar nerve cells. The axons are directed toward the base of the cochlea (figs. 33, 34) and constitute the cochlear nerve. Reaching the internal acoustic meatus, the cochlear nerve travels with the vestibular nerve (both being divisions of the eighth or acoustic nerve), and the facial nerve, to the ponto-

→

Figure 34. STRUCTURE OF COCHLEA AND ORGAN OF CORTI

A is an axial cut showing one-half of the cochlea, surrounded by bone, with the cochlear nerve in the axis sending out its fibers in a spiral manner. B is a single turn of the cochlea, with its three spiral chambers, and the organ of Corti. C shows the floor and wall of the cochlear duct with the tectorial membrane in place. D is a stereogram of a short sector of the organ of Corti.

Bas. Memb.	Basilar membrane
B.C.	Border cell
C. H.	Cells of Hensen
Ha.	Auditory "hairs"
H. P.	Heads of pillars

I. H. C.	Inner hair cells
I. P.	Inner pillars
I. P.	Inner phalangeal cells
I. T.	Inner tunnel
N. F.	Fibers of cochlear nerve
O. H. C.	Outer hair cells
O. P.	Outer pillars
O. P. C.	Outer phalangeal cells
O. Ph.	Outer phalanges
O. T.	Outer tunnel
Ret. memb.	Reticular membrane
Spir.	Spiral ganglion
S. T.	Scala tympani
S V.	Scala vestibuli

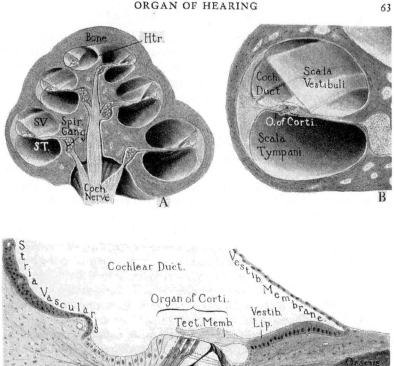

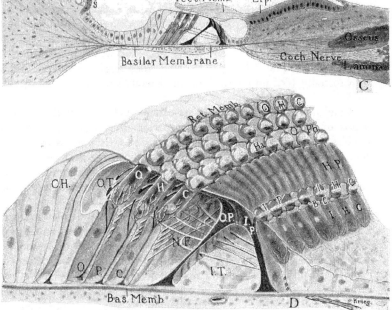

bulbar junction (frontispiece; 261: V L). (The auditory system is rendered in Prussian blue in the diachrome.) Here, on the surface of the restiform body, are plastered the conjoined dorsal and ventral cochlear nuclei like a flattened dumbbell (208, 207: IV; figs. 29, 49: actually they are lateral to the restiform body, 242: ·V, and not medial).

AUDITORY PATHWAY. The cochlear nuclei contain secondary cell bodies on which terminate the primary fibers of the cochlear nerve (fig. 49). The dorsal and ventral cochlear nuclei are not identical in significance, or in cell type. The dorsal cochlear nucleus is apparently concerned with transmission of the reflex and affective components of hearing, while the ventral cochlear nucleus relays the discriminative aspects, the pure tones. The character of the synapses indicate this. In the dorsal nucleus the endings are more diffuse and the cells send their dendrites over a wider space; in the ventral nucleus a primary fiber clasps a cell body like a hand holding a ball (fig. 1: G).

In the manner of secondary sensory fibers generally, the axons of both cochlear nuclei stream medially to gain the opposite side of the brain stem. Some from the dorsal cochlear nucleus run over the top of the restiform body as the acoustic stria (fig. 49). Those from the ventral cochlear nucleus bend under the restiform body as the beginning of the trapezoid fibers (fig. 49: Trap.). The acoustic stria fibers drop ventrally through the reticular substance; the trapezoid fibers continue medially, and the two sets join and form the definitive trapezoid fibers (209: IV M, figs. 29, 49, 50). They migrate upward in the pons as they cross and run laterally in the opposite side, spreading out over a considerable area, and weaving across the medial lemniscus (175, 176: III; fig. 29). When they reach the opposite side, they turn rostrally (*i.e.*, up the stem) as the lateral lemniscus (210: IV; figs. 29, 50, 51).

Cradled on the trapezoid body are certain nuclear accumulations forming a part of the auditory system. The superior olive (figs. 29, 49: S. O.) resembles in cell type the dorsal cochlear nucleus. It receives primary auditory fibers, and in part sends its secondary fibers up the lateral lemniscus of the same side. This means that each ear is represented on both sides of the brain and is the reason that complete deafness is never a symptom of localized brain injury. The superior olive is also a reflex center. It sends to the M. L. F., making connections with all the eye muscles and cervical cord, which turn eyes and head in response to a sound; it connects to the facial nucleus adjacent, causing eye closure in response to a loud sound. This reflex can be used to determine whether an infant is deaf. Connections with facial and trigeminal motor nuclei cause reflex contraction of stapedius and tensor tympani muscles of the middle ear, which damp the vibrations of stapes and malleus during a loud sound. It also sends fibers out to the cochlea, but for what purpose it is not clear. The nucleus of the trapezoid body, medial to the superior olive, is more diffuse. Its cells and synapses resemble those of the ventral cochlear nucleus, and it receives primary fibers. Apparently it functions merely as a displaced portion of that nucleus. Along the course of the lateral lemniscus are two small nuclei of the lateral lemniscus (fig. 29: NL). They send reflex fibers directly across the median plane, which turn and descend.

On reaching the midbrain, the lateral lemniscus breaks up in the large inferior colliculus (81: I; 211: IV; figs. 29, 32, 51), a conspicuous pair of elevations on the dorsal surface of the midbrain. They are much larger in lower mammals (rat, 211: IV). Its cells are small and scattered about without apparent organization or tonal localization. Reflex connections are made with the brain stem, some descending with the tectospinal tract (141: II). Auditory connections to the cerebellar cortex pass back along the lateral lemniscus. Many of the lateral lemniscus fibers do not synapse in the inferior colliculus, but bypass it to continue toward the thalamus, and of those that do, many of the tertiary fibers continue with the secondary fibers. This combined bundle is the brachium of the inferior colliculus (212: IV; figs. 39, 51). It reaches the globular medial geniculate body (213: IV; figs. 37, 54), the auditory nucleus of the thalamus. The sole function of this nucleus seems to be a way station to the cerebral cortex. Some experiments indicate tonal localization is lacking. The auditory radiation (214: IV; fig. 54), of thalamocortical fibers, leaves the thalamus by running laterally and joins the internal capsule in its infralenticular portion. Here the fibers are laterally directed and enter the temporal lobe (214: IV L). They end in area 41 of the cortex, the auditory receptive area (41: VI L; fig. 54) located on the hidden upper surface of the temporal lobe; in other words, on the lower bank of the deep lateral fissure (278: VI L). Area 41 reaches the lateral surface barely, or not at all. The handling of auditory stimuli by the cortex will be taken up in the next chapter (p. 100).

VISION

It is through sight that bony fishes, amphibia, reptiles, and, among mammals, the primates (monkeys, apes, man) obtain food and avoid enemies. Many animals have sharper eyes than we, but we make the most of our vision, for it is the most widespread source of our instruction and culture. The auditory system deals with common and traditional knowledge, group instruction, and incidental information; but vision is used to acquire information on whatever is complex, recondite, uncommon, or new. The auditory region has a stronger link to memory, personality integration and emotion, but the eye with the intellect.

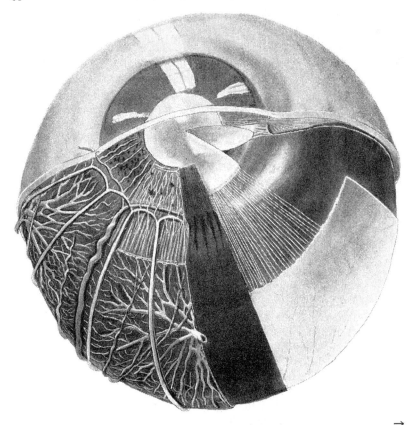

→

Figure 35. EYEBALL DISSECTED BY LAYERS

The upper half is intact. The left side of the lower half shows the chorioid layer with its arteries, veins, nerves, and muscles; the right side shows, successively, the pigmented layer of the retina; the suspensory ligament, the optic part of the retina, and the vitreous humor.

The arteries may be grouped as anterior ciliary, entering near the sclero-corneal junction; and posterior ciliary, entering around the optic nerve. The latter group are short or long. The short ramify in the chorioid, the long continue to the edge of the iris, where they anastomose with the anterior ciliary to form the arterial circle. The veins, very different in arrangement, converge radially on the surface of the chorioid to form four trunks, leaving the eyeball near the equator, as the vorticose veins.

The ciliary nerves, derived from the ciliary ganglion, enter around the optic nerve, run as longitude lines as far as the iridal edge, where they anastomose to form the ciliary ring, and supply the ciliary and iridal muscles.

(Redrawn from Krieg: Functional Neuroanatomy; McGraw-Hill Book Co., N.Y. 2nd edition, 1953)

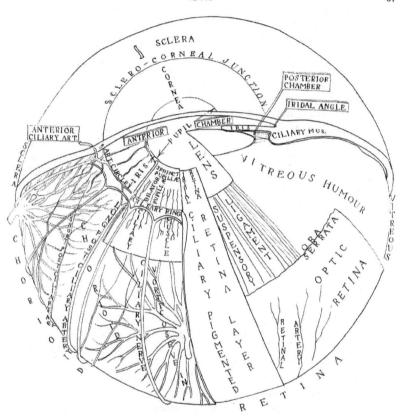

Figure 35A. KEY TO ILLUSTRATION ON OPPOSITE PAGE

THE EYE in its smooth rotundity seems a simple organ, especially when it is seen to be thin-walled and hollow, but in reality it is a complex and marvelously designed mechanism. Space requirements forbid an adequate description, but reference is here made to figure 35 and 35A. The essential features of the eye are a lens mechanism formed by the curved front of the eye plus the lens inside, which together focus an image on the light-sensitive back wall of the eye, the retina. The eyeball is composed conventionally of three layers. The outer, the sclera, is of connective tissue, tough as leather. Its anterior part, the cornea, is of sharper curvature and is transparent. The middle layer, the chorioid, carries the blood vessels and nerves, but not the optic nerve. It also contains the ciliary muscle, which changes the shape and focal length of the lens by pulling on the edges of the

transparent envelope that contains it. In front the chorioid layer is separated from the cornea to form the iris, a perforated and adjustable light-
diaphragm. In the iris are circularly arranged smooth muscle fibers, the
sphincter pupillae, and radial fibers, the dilator pupillae. The inner layer,
the retina, is the one in which we are most interested here, for its posterior
half contains the entire light-receptive mechanism. The retina continues
anteriorly over the chorioid muscle and the iris, but there it carries a black
pigment and is not light-sensitive.

The optic retina looks like a piece of tissue paper, with a few thin radiating red lines marked on it — the retinal vessels from the ophthalmic
branch of the internal carotid artery. Yet within this tissue is all the remarkable mechanism about to be described. Microscopic sections (fig. 36, left)
show three layers of nuclei of cells and three layers of synapses; there are
three orders of neurons in the retina (see fig. 36 for all details). Furthest
toward the outside of the eye is the receptor layer composed of the cones.
The light must pass through the other layers before reaching these, the only
light-sensitive elements. The rods (r) are shaped like a cattail on its stalk,
the cones (c) are similar, but thicker at the base. The rods are about ten
times as numerous as the cones, except at the spot at the middle of the retina
where the central ray of light focuses, the macula, and this is composed
entirely of cones. This is the most sensitive spot of the retina, the only one
good for reading and other discriminative work. Many lines of evidence
conspire to prove that the rods are to make vision possible at all in dim light,
while the cones are for detailed and color vision in bright light. Many rods
converge to one optic nerve fiber, but each cone has its private fiber to the
brain. The rods and cones are only parts of their respective neurons. The
nuclei are located in the outer nuclear layer. The remainder of the rod neuron
is an extremely tenuous fiber which extends to the outer plexiform layer,
and ends by a single knob. Its nucleus is like one bead on a string. The axon
of the cone neuron is stockier, straight, and ends like a little hand with fingers
extended. Its nucleus is a bulge joined to the base of the cone.

Both types synapse with the bipolar cells. Their nuclei are in the inner
nuclear layer, their dendrites in the outer plexiform layer, and their axons
in the inner plexiform layer. The rod bipolar (cb, db) dendrites fan out to
gather from a number of rod terminals. The cone bipolar (b) dendrites are
like a hand whose fingers join with the fingers on the terminals of a single
cone.

The third neuron is the ganglion cell. They appear more like proper
neurons. Their cell bodies are in the ganglionic layer, their dendrites synapse
with the bipolar cells in the inner plexiform layer, their axons run along the
inner surface of the retina, converge, and plunge through the back of the

eye to form the optic nerve. The dendrites of the rod ganglion cells (pg) branch out like an elaborate chandelier and contact the terminals of many rod bipolars. Other rod bipolars send vine-like terminals among the ganglionic cell bodies. The cone ganglionic cells are smaller. They send out a single dendrite which clasps hands with a single bipolar terminal. To gain an even greater gathering power to the rod system there are horizontal cells (h) in the outer and inner plexiform layers whose dendrites collect from a large number of rod and cone terminals and whose axons apparently dis-

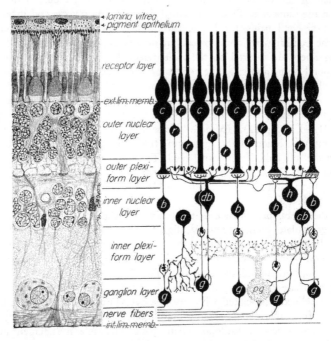

Figure 36. ON THE LEFT, A STRIP OF A MICROSCOPIC SECTION OF THE RETINA; ON THE RIGHT, A DIAGRAMMATIC INTERPRETATION OF THE RETINAL NEURONS

(From Walls: The Vertebrate Eye. Cranbrook Institute of Science, Bloomfield Hills, Mich. 1942)

charge on rod bipolar neurons. The diffuseness of the rod system and the specificity of the cone system are evident from their connections in the retina.

OPTIC PATHWAY. At the macula all the overlying layers are parted back to permit the maximum preciseness of the image. Where the converg-

ing ganglionic axons leave the eye, at the optic disc, there are no visual elements, and, consequently, a blind spot exists. On leaving the back of the eye
to form the optic nerve, a myelin sheath is acquired. The optic nerves run
backward in the orbits (fig. 30), enter the brain cavity (116: I; 215: IV)
by the optic foramen, and converge to form a partial crossing at the optic
chiasma (216: IV; 77: I; frontispiece). Only the fibers from the medial or
nasal half of the retina cross, those from the lateral or temporal half continue on the same side. Since the retinal fields are the reverse and inversion
of the peripheral fields, the lateral parts of the entire visual picture cross,
while the central part is uncrossed. In tumors of the pituitary gland, the
crossing fibers are injured, so the patient feels he is wearing side blinders
(bitemporal hemianopsia). The object of this crossing is to combine the
fibers which are carrying the image from one side of the body, for it is
obvious that the lateral or temporal side of one eye sees the same image that
the medial or nasal half of the other eye sees. The new bundle is the optic
tract (217: IV; fig. 52). (Middle blue is used for the optic system in the
diachrome.) A lesion of the right optic tract will produce complete blindness on the left half of the visual field. The optic tracts of either side run
horizontally backward, forming a V to enclose the hypothalamus, and then
grasping the peduncles of the midbrain like a thumb and forefinger around
an arm (217: IV L).

At the lateral prominence of the brain stem the optic tracts end in the
lateral geniculate bodies (249: V; figs. 37, 53, 54), large helmet-shaped
nuclei. In section they are seen to be composed of alternate thick laminae
of cells and thin sheets of fibers, like a six-layered cake and its filling, somewhat humped in the middle (fig. 54). The fibers derived from either eye
occupy alternate laminae and discharge into the layer above them, so alternate
layers of cells are associated with any one eye. Throughout the optic tract
the fibers from the macula occupy the middle swath, and because they are
so numerous, occupy a disproportionate space. The same arrangement holds
in the lateral geniculate, but the macular fibers occupy an even larger area
proportionately of the central sector; in fact, they are responsible for the
humped shape. The fibers from the lateral edge of the visual field occupy
the medial and lateral corners.

The lateral geniculate body is seemingly below and lateral to the thalamus, but in reality is a part of it, since its sole purpose is to project optic
fibers to the cortex. The optic radiation (geniculocalcarine tract) spreads out
from its dorsal, lateral and posterior aspects like smoke from it blown by
a breeze from the rostromedial direction (250: V; fig. 54). It passes backward to the occipital cortex, but in order to reach it, must clear the lateral
ventricle at its deepest curvature (99: I L; fig. 55). The lowest fibers are

pulled considerably out of course in a forward direction by the inferior horn of the lateral ventricle (see 251: V L). In injury to the temporal lobe, this group may be selectively damaged, causing blindness of the upper temporal quadrant of the opposite side.

The optic radiation is flattened against the lateral surface of the ventricle (232: V L), forming much of the external sagittal stratum, a broad vertical band of fibers (250-254: V; figs. 54, 55). Within the radiation the optic fibers are in a similar order to what they have been: the macular ones in the middle (254), those from the main part of the retina (binocular) above and below (253), and at upper and lower edges are those from the part of the field so lateral that its image falls only on one eye (monocular, 252). The image on the retina is inverted, and this inversion is continued in the optic radiation.

Half-way back, as soon as the posterior horn of the lateral ventricle (98: I L) can be cleared, fibers from upper and lower edges peel off and enter the cerebral cortex above and below the deep calcarine sulcus (I: 64). This continues to the occipital pole, deeper and deeper fibers coming to their termination in the cortex around the calcarine sulcus. This means that the monocular quadrants are represented furthest forward in the calcarine cortex, the binocular quadrants in the middle, leaving the macular fibers to end in the occipital pole cortex, beyond the limits of the calcarine sulcus. The visual receptive cortex is area 17 (17: I, VI). On the medial surface it tapers to a sharp point, but actually it is folded deep into the cerebral hemisphere; on the occipital pole it broadens out like a tennis racket. There is a strict point-for-point localization of the visual field, with the lower half above the sulcus, and the upper half below it.

VISUAL REFLEXES. Some visual fibers bypass the lateral geniculate, continue their course around the brain stem and enter the superior colliculus below an overlying layer of cells. This is the brachium of the superior colliculus (figs. 51, 54: B. S. C.). The superior colliculus (80: I; II M; figs. 39, 51) is the optic reflex center of the brain stem. In interpreting its function, we should remember that in lower vertebrates, as the optic tectum, it had the most elaborated structure, and was the most widespread in its connections in the entire brain (fig. 13). When optic perception was shifted to the cortex, a rich supply of reflex connections was retained. Such are (1) to the nucleus of Edinger-Westphal, enabling the pupil to respond to light; (2) to the eye-muscle nuclei; (3) to motor nuclei of the head and neck (tectobulbar tract); and (4) to spinal motor nuclei (tectospinal tract). The last two enable immediate bodily response to sudden appearances. The fibers drop ventrally from the deep cell layer of the colliculus, decussate and pass down the brain stem (141: II; figs. 51, 50, 49), just below the MLF (95: I), and with it drop to a ventral position (fig. 48) in the cord (figs. 5, 7: Te.-Sp.).

BODILY SENSATION

MEDIAL LEMNISCUS. In our study of the spinal cord, we found that two main groups of fibers ascend to the cerebrum: the spinothalamics, which are secondary fibers and proceed without interruption to the thalamus (181: III); and the fibers of the dorsal funiculus, which are primary (p. 18). We shall now trace these. In lower levels of the bulb dorsally are two large, elongated nuclei, nucleus gracilis, medially; and the nucleus cuneatus, laterally (178, 179: III; figs. 29, 32, 48). These are the much displaced secondary nuclei in which the primary fibers of the dorsal funiculi synapse. Recalling the precise topology in this tract (fig. 7), the gracilis is from the lower limb, the cuneatus from the upper limb. A glance at figure 48 will show that these nuclei are properly located for the sequence of nerve components, dorsolateral to the visceral sensory component. Following the general rule for secondary sensory fibers, the axons from these nuclei arch around the other nerve nuclei as internal arcuate fibers (177: III; figs. 32, 48), cross to the other side and turn rostrally, this time next to the midplane. They form the medial lemniscus, which is a sagittal lamina in the bulb (176: III; figs. 32, 49), becomes flattened transversely in the pons (175: III; figs. 29, 32, 50), migrates to the lateral surface in the midbrain (174: III; figs. 29, 32, 51), then plunges into the thalamus (174: III). While it is migrating laterally in the pons, the spinothalamic tract (181: III; figs. 48-50) becomes incorporated into its lateral aspect, and, after a brief flirtation with the lateral lemniscus, becomes completely merged with the medial lemniscus.

TRIGEMINAL SYSTEM. We have not yet followed the somesthetic sensory fibers from the face. Virtually all are carried by the sensory division of the trigeminal nerve, which is much larger than the masticatory division. Leaving the lateral aspect of the pons (257: V L; fig. 30), the trigeminal nerve, the largest cranial nerve, becomes strap-shaped and soon joins the large semilunar ganglion (258: V L), inside the skull. This is a preamble to its division into three branches: the ophthalmic, which enters the orbit and supplies the eye and forehead; the maxillary, which supplies the upper jaw, nose, and palate; and the mandibular, carrying sensation from lower jaw and tongue. The details of the distribution and branches of ramification are shown in figure 30.

The widest possible variety of somesthetic modalities are found in the sensory field of the trigeminal nerve. The cornea and teeth feel only pain, the nostrils are sensitive to tickling, the lips to light touch; the tongue has excellent tactile localization; and the vibrissae or "whiskers" of most mammals have elaborate sensory endings (fig. 2: A). These are protective or informative and furnish the basis for a number of reflexes such as blinking, sneezing, suckling, chewing.

In the painful affliction known as trigeminal neuralgia, the sensory field of the trigeminal is outlined by attacks of excruciating pain. It is relieved by section of the trigeminal root, which does not regenerate, and which is separate from the motor root.

The trigeminal sensory root has a long course dorsally through the pons (234: V M; figs. 28, 29, 50). Near the ventrolateral angle of the fourth ventricle it seemingly turns bodily down the brain stem; in reality, many of the fibers bifurcate. The upper rami end in an ovoid main sensory trigeminal nucleus (236: V M; figs. 28, 29). The lower rami form the massive spinal trigeminal tract (or "spinal V tract"), which descends to the spinal cord (235: V; figs. 28, 29). In the pons it is between vestibular and facial nerve roots, as it should be in the sequence of nerve components (235: fig. 50); in the bulb it migrates to the surface (235: figs. 49, 48) and finally joins Lissauer's tract of the cord. Just medial to the tract all along the way is the secondary nucleus. Actually, there are two: the gelatinosus, continuous with the nucleus of the same name in the cord, and the spongiosus, continuous with the body of the sensory cell column of the cord. The gelatinosus relays nociceptive stimuli, the spongiosus discriminative. The gelatinosus is prominent below and absent above, but the spongiosus runs all the way (figs. 48, 49, 50). The spongiosus is made up of several subnuclei. In the root the mandibular fibers are dorsal, the maxillary fibers in the middle, and the ophthalmic fibers ventral. The mandibular fibers peter out first, the maxillaries half way down, but the ophthalmics run all the way. When the spinal V tract is sectioned at operation in the upper bulb, above the gelatinosus, pain is lost from the entire face. There is not space here to discuss the numerous reflex connections of the trigeminal; for this, Krieg's "Functional Neuroanatomy" may be consulted.

The secondary trigeminal fibers cross directly, but are scattered all along the lengths of pons and bulb and mixed with other fibers of the reticular substance, so are not evident. The nociceptive fibers run far lateral and ascend in company with the spinothalamic tracts, the discriminative fibers join the medial lemniscus as the trigeminal lemniscus. In addition, a dorsal secondary trigeminal tract is classically recognized, which emanates from the main trigeminal nucleus and pursues an independent course through the middle of the reticular substance.

There is another sensory division of the trigeminal. This one breaks all the rules. Its cells, although large and round like other primary sensory cell bodies, are located within the midbrain ventrolateral to the aqueduct. This mesencephalic trigeminal (or "V") root (237: V M; figs. 28, 29, 50, 51) is proprioceptive for the act of mastication, sending sensory endings to the jaw muscles and sockets of the teeth. Each fiber sends off a collateral to the masticator nucleus. How the eye muscles, muscles of expression, or tongue and larynx muscles get their proprioceptive supply is one of the classic problems of neuroanatomy, but it is not mediated through the mesencephalic V root.

SOMESTHETIC THALAMUS. As detailed above, all of the secondary somesthetic fibers combine into one large tract at the upper end of the midbrain, the augmented medial lemniscus (174: III; figs. 39, 51). In this tract, all modalities are combined, but the segments are arranged in an orderly manner. The fibers from the head are most medial, then follow, successively, upper limb, trunk, lower limb. It then enters immediately the somesthetic nuclei of the thalamus, where it terminates. There are two of these, one for the head fibers, the arcuate nucleus (ventralis posterior medialis, VPM) (149: II; figs. 37, 38, 39, 54); and the somesthetic nucleus (ventralis posterior lateralis, VPL) (148: II; fig. 48, 54). The somesthetic nucleus is placed in that

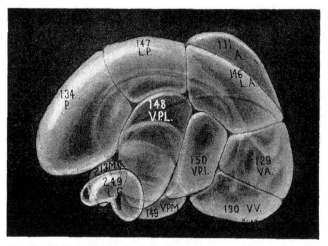

Figure 37. BUBBLE RECONSTRUCTION OF THALAMIC NUCLEI OF RIGHT SIDE, LATERAL ASPECT

(For legend see Figure 38)

part of the thalamus indicated by its Latin name. It is comma-shaped, the arcuate nucleus fitting into its cavity. The arcuate nucleus is similar but smaller, and is medial. Into the arcuate, in turn fits the rounded central nucleus (137: same figures), which is no longer considered to receive head somesthetics (see p. 125).

The arcuate and somesthetic nuclei show the arrangement of the parts of the body projected upon them by the medial lemniscus, head medial, feet lateral; but the cells seem scattered and without any particular order or grain. These nuclei are way stations on the somesthetic pathway to the cortex. The contained cells send axons laterally and dorsally, as the somesthetic radiation (157: III; figs. 39, 54). It joins the internal capsule (157: IV F), whose fibers are running nearly vertically at this point, and fans out (3: IV L) to be distributed to the cerebral cortex along the posteroinferior wall of the central sulcus (3, 1: VI; fig. 54) to cortical areas 3, and to a lesser degree, 1, the somesthetic receptive areas. Within the radiation and on the cortex, a very orderly arrangement is retained, so that the opposite half of the body is represented, but upside down (see VII). The arrangement and distribution of the somesthetic fibers in the cortex will be discussed in the next chapter (p. 93).

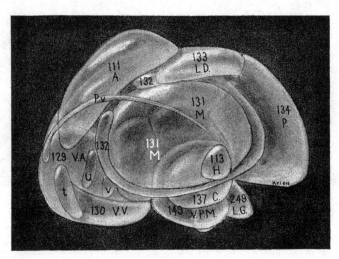

Figure 38. BUBBLE RECONSTRUCTION OF THALAMIC NUCLEI OF RIGHT SIDE, MEDIAL ASPECT

KEY TO FIGURES 37, 38, 39

111. Anterior nucleus (A.)
113. Habenula (H)
129. Nucleus ventralis anterior (V. A.)
130. Nucleus ventralis (V. V.); ventralis lateralis (V. L.)
131. Nucleus medialis (M.)
132. Intralaminar nuclei
133. Nucleus lateralis dorsalis (L. D.)
134. Pulvinar (P.)
137. Nucleus centralis (of Luys) (C)
146. Nucleus lateralis anterior (L. A.)
147. Nucleus lateralis posterior (L. P.)
148. Nucleus ventralis posterior lateralis (V.P.L.) or somesthetic nucleus
149. Nucleus ventralis posterior medialis (V. P. M. or arcuate nucleus)
150. Nucleus ventralis posterior inferior (V. P. I.)
213. Medial geniculate body (M. G.)
249. Lateral geniculate body (L. G.)
Pv. Nucleus paraventricularis
t. Nucleus reuniens
u. Nucleus centrum medianum
v. Nucleus submedius

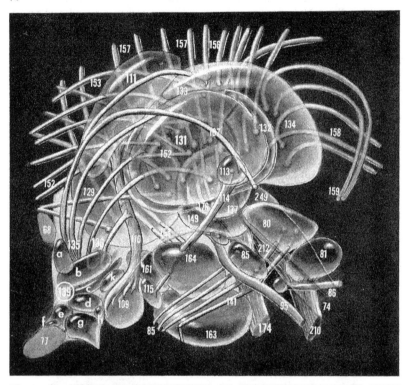

Figure 39. BUBBLE RECONSTRUCTION OF DIENCEPHALIC NUCLEI, THEIR
CONNECTIONS AND CERTAIN STRUCTURES OF MIDBRAIN.
RIGHT SIDE, MEDIAL ASPECT

The thalamic nuclei are identical with
those of fig. 38. For other structures see list
below.
68. Anterior commissure
77. Optic chiasma
80. Superior colliculus
81. Inferior colliculus
85. Oculomotor nucleus
85n. Emergent fibers of oculomotor nerve
86. Trochlear nerve
95. Medial longitudinal fasciculus
107. Stria medullaris
109. Mammillary body
110. Mammillothalamic tract
114. Habenulopeduncular tract
115. Interpeduncular nucleus
139. Hypothalamic nuclei
    a. Nucleus preopticus
    b. Nucleus magnocellularis hypo-
      thalami

c. Nucleus filiformis hypothalami
d. Nucleus ventromedialis hypothalami
e. Nucleus anterior hypothalami
f. Nucleus suprachiasmaticus
g. Nucleus arcuatus hypothalami
h. Nucleus premammillaris
k. Nucleus dorsomedialis hypothalami
141. Tectospinal tract
152. Anterior thalamic radiation
153. Subcallosal anterior thalamic radiation
156. Projection of nucleus lateralis posterior
157. Somesthetic radiation
158. Posterior thalamic radiation
159. Arnold's thalamic fasciculus
161. Subthalamic nucleus
163. Substantia nigra
164. Red nucleus
165. Brachium conjunctivum
174. Medial lemniscus
210. Lateral lemniscus
212. Brachium of inferior colliculus

Group and Nucleus	Desig-nation	Location	Afferents	Efferents
HABENULA	113 H	Postero-dorso-medial corner	Stria medullaris	Habenulopeduncularis
MIDLINE GROUP (6 small nuclei)		Around medialis at midplane	Ansa peduncularis	Mostly noncortical Dorsal long. fasc.
ANTERIOR	111 A	Rostro-dorso-medial elevation	Mammilloth-alamic tr.	Cingulate region: Areas 23, 24, 29
MEDIAL	131 M	Inner, medial ovoid	Unknown	Prefrontal cortex
INTRALAMINAR GROUP (3 small nuclei)	132	Thin cup outside medialis	Ascending activator neurons	Diffuse to cortex?
POSTERIOR GROUP Parafascicular		Behind medial n. Medial to central Nuc.	Ascending activator neurons. Pallidus	Basal frontal cortex
Central (Luys)	137 C	Below medial, medial to somesthetic	Ascending activator neurons. Pallidus	Pallidus
VENTRAL GROUP Ventralis anterior	129 VA	Rostral end	Ascending activator neurons? Pallidus	Pallidus. Prefrontal cortex
Ventralis ventralis (or ventralis lateralis)	130 VV	Rostral end, below	Brachium conjunctivum	Areas 4, 6
Ventralis posterior lateralis (somesthetic)	148 VPL	Lateral to and below medial	Medial lemniscus, spinothal.	Areas 3, 1
Ventralis posterior medialis (arcuate)	149 VPM	Medial to VPL	Secondary trigeminal tracts	Areas 3, 1; lower third
Ventralis posterior inferior	150 VPI	Lateral to medial; between VV & VPL	Med. lemniscus indirectly	Areas 4, 6, 8
LATERAL GROUP Lateralis dorsalis	133 LD	Dorsal to medial	Pallidus	Pallidus Noncortical
Lateralis anterior and posterior	146 147	Dorsolat. to medial, dorsal to VP	Unknown	Areas 6, 8, 7, 40
Pulvinar	134 P	Posterior bulge; massive	Unknown	Areas 39, 40, 37, 19, temporal lobe
GENICULATE BODIES Medial geniculate	213	Posterior, inferior, lateral	Auditory tract	Area 41
Lateral geniculate	249	Lateral to medial geniculate	Optic tract	Area 17
RETICULAR		Thin lateral shell for thalamus	Thalamic activator neurons	Diffuse to cortex

SMELL

We have now followed all of the nerve connections which enter the central nervous system to their termination except the olfactory component. They tend to follow certain principles of arrangement of primary, secondary and tertiary fibers, except for the peripheral parts of the visual system (and the mesencephalic V). The olfactory system is a law unto itself. Entering the rostral end of the cerebrum directly, it has no long, ascending tracts, or wayside connections. It immediately plants its roots in the loose, rich soil of the cerebrum, and holds fast through the vertebrate phylum to whatever it has claimed.

The olfactory epithelium in the upper part of the nasal cavity is a thick, simple columnar epithelium containing some olfactory cells, scarcely differentiated in appearance from the other epithelial cells, but their basal part continues in the subepithelial tissue as an unmyelinated nerve fiber. Collecting into threads, these perforate the bone and immediately enter the olfactory bulb (103: I L; fig. 30, frontispiece), lying directly over. The olfactory bulb is a part of the cerebrum, not a nerve, and has a structure basically similar to the cerebral cortex. The axon ends synapse with the dendritic terminals of the secondary or mitral cells by a particularly intimate connection, the olfactory glomerulus (fig. 1: H). In the bulb are very numerous granule cells which perhaps step up the intensity of the stimulus.

The axons of the mitral cells run back in the olfactory tract, which is separate, like a nerve (104, violet: I L; frontispiece). In the rat the olfactory bulb is directly continuous with the cerebrum because the growth of the frontal lobe has not distorted the direct relationships (fig. 17). Comparison with the rat will be particularly interesting when studying the olfactory system. Reaching the base of the hemisphere, the olfactory tract diverges into three bundles, just as in the salamander (p. 26): (1) the medial olfactory stria (105) passes to the septum (66: I M); (2) the intermediate olfactory stria ends immediately in the olfactory tubercle (frontispiece); (3) the lateral olfactory tract makes a sharp angle laterally and ends in amygdala, hippocampus and pyriform cortex. All of these terminals are part of the cerebral cortex, or a modification of it, so will be considered in the next chapter. (Pp. 79-82, 117-118)

Chapter Six

CORTICAL ANALYSIS AND SYNTHESIS

WHAT IS THE CEREBRAL CORTEX?

IMPLER GRADES OF CORTEX. The organized system of neurons that covers the cerebrum, and thus most of the surface of the human brain, namely, the cerebral cortex, is the most highly elaborated part of the brain and of the body; the least understood, but the most significant. It is the most complex structure in the world we know, and probably the most important. To gain any insight into the mechanism and organization of the neocortex, which accounts for nearly all the cortex in man, we cannot do better than study the more primitive types of cortex which have remained in parts of the human brain, as though charmed to immobility eons ago.

The neuronal pattern of the cerebral hemisphere of the salamander (p. 29) is simple and uniform. Cell bodies in a layer next to the ventricle send radiating clusters of dendrites outward into the overlying zone of afferent fibers, and send their axons toward the ventricle. These continue to lower levels, or return to the outer layers. One type has been degraded from even this low level of organization: the septum. Due to the fact that the medial olfactory tract leaves the surface and wanders among the cell bodies, their dendrites have lost orientation and are scattered as in an unorganized nucleus. The septum (66: I M) receives the medial olfactory stria (105: I L) and sends out a few fibers which perforate the callosum and run backward over its top (fornix longus) to the hippocampus.

In the hippocampal formation are two primitive types of cortex. In the human brain the hippocampus (230: V; figs. 53, 54) runs along the inferior horn of the lateral ventricle (232), with a "paw" at the forward end, the pes hippocampi (231: V; fig. 53). In the rat (230: fig. 22), it is a flattened roll, relatively much larger in the posterior part of the cerebrum. It is curved in cross-section (fig. 40: A) like the letter S; the dentate gyrus forms one

[79]

limb and type, the hippocampus proper forming the other. The neuronal pattern of the dentate gyrus is like that of the salamander cortex: a layer of cell bodies radiating dendrites outward into a layer of fiber ramifications, and sending axons basally (fig. 40: A). The hippocampus proper is a step more complex, for its single layer of cells has a spray of dendrites below as well as one above, which receive afferents of different nature (fig. 40: A). The axons pass basally and gather into the fornix (67: I; figs. 52-55), which curves in nearly a complete circle to end in the mammillary bodies (109: I; also rat, I L, V M). The neurons of the dentate gyrus may be taken as representing layer ii of the definitive cortex, those typical of the hippocampus proper representing layer v. It is evident that afferent streams in cortical strata are the determining factor in dendritic arborization patterns.

Primary olfactory cortex (pyriform cortex, or area 51) is the next stage. In man it has only a small representation, but in the rat (51: rat, I, VI) it is extensive, as it includes the olfactory bulb (103) and the basal part of the cerebral cortex (51). Its outer fiber lamina, layer i (fig. 40: B) is provided by the olfactory stria. The cells of the outer layer (layer ii) correspond to those of the dentate gyrus and are very dense. Their dendrites ramify in the olfactory stria. Below is a thicker zone of more scattered cells, interspersed with many fibers of extraneous origin, forming the undifferentiated layers iii-vi. While the single cell type of the dentate gyrus gathered excitation from only one fiber layer, and the characteristic cell type of the hippocampus gathered impulses from laminae both above and below, this cortex contains separate cell layers for superficial and deep currents, and the means of integrating them.

A higher cortical type is seen in the secondary olfactory or entorhinal cortex, areas 28, and 35, forming the hippocampal *convolution* and uncus (28, 35: fig. 42A, p. 105, figs. 52-54). In the rat it is further back and relatively large (28: fig. 22). It is evident from the diagram (fig. 40: C) that this kind of cortex has more cellular differentiation than the preceding. The four conspicuous dendritic plexuses shown correspond to the four kinds of afferents, and the 37 neuron types shown indicate the variety of handling possible. The numbering of layers indicates that the diffuse layer iii-vi has become differentiated by the afferents, but that a separate layer iv is still lacking. The great achievement here is the pyramidal cell (numbered 6-19), which can send out dendrites to any combination of layers, integrate the

→

Figure 40. THE SIMPLER GRADES OF CORTEX

A. The dentate gyrus and the hippocampus
B. Primary olfactory cortex
C. Secondary olfactory cortex
For explanation see text.

(B from Cajal: *Histologie du Systeme Nerveux, v. 2, Maloine, Paris. 1911. C from Lorente de No, in Fulton: Physiology of Nervous System, Oxford University Press, New York, 3rd edition, 1949*)

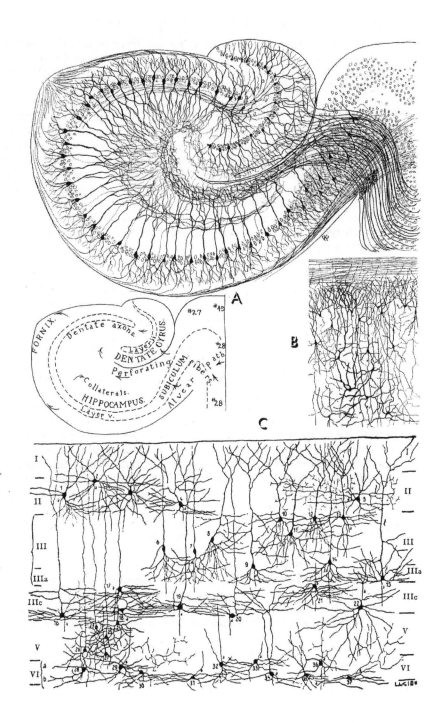

A

FORNIX.
Dentate axons
Layer II.
DENTATE GYRUS.
Perforating fibers path
Collaterals.
HIPPOCAMPUS.
Layer V.
SUBICULUM.
Alvear
#27   #49
#28
#28

B

C

I   I

II   II

III   III

IIIa   IIIa

IIIc   IIIc

V   V

VI {a   VI
   {b

L.K.18

impulses and distribute them selectively to any layer, or send them outside the cortex, by means of its axon and collaterals. By virtue of the pyramidal cell, the cortex may be regarded as divided into a great number of vertical prismatic functional units. This innovation proved so useful that it was retained as the principal feature of the highest cortical type, neocortex.

MECHANISM OF THE NEOCORTEX. The cortical types described are strongly connected with the olfactory system and constitute the cortical part of the rhinencephalon. The remainder of the cortex, the neocortex, is connected directly or indirectly with the thalamus by fibers which end in a newly developed lamina, layer iv, and this exerts a profound influence on the arrangement of neurons. Its cell types and their distribution are shown in figure 41. Because layer iv cells are pale and very small, they form a plane of reference for identifying and designating the other layers (fig. 42: iv in each type). For that reason it is called the granular layer, hence the layers i-iii are supragranular and the layers v and vi are infragranular. These differences extend into the functional realm also, supra- and infragranular layers contain different types of neuron circuits. The thalamic axons terminate in a dense plexus within layer iv (fig. 41: a), and the lowest part of layer iii (iii-c) (fig. 41: b). The cells of iv act only as distributive cells within the cortex, their axons and collateral branches turn outward to ii and iii and downward into v and vi (fig. 41: e-g). The infragranular layers may be regarded as completing the shorter intracortical circuit, when the dendrites there do not pass external to the zone of thalamic afferents, and the axons leave the cortex. When the thalamic impulses are carried out to the supragranular layers, the axons from cells in ii and iii drop down to the infragranular layers and end in contact with cells there, which in turn send out efferent axons, we have the longer intracortical circuits. For some reason there is a strong tendency of dendrites of all layers to reach out to layer i, which has no cells of its own except a few with horizontally oriented axons and dendrites, but which does contain horizontal axons of often distant origin. Possibly these form a widespread diffuse suppressor system which inhibits discharge from all points save one at any instant. There are certain large pyramidal cells in layer v (fig. 42: K), they have widespread dendritic ramifications in layers v and i (fig. 41: n, o), and send out a stout axon. These are the projection neurons to subcortical structures, characteristically to motor nuclei. The three other types of axon course are: (1) corticothalamic, back to the thalamic nuclei which have excited them; (2) callosal, to a similar unit of cortex on the opposite side; and (3) associational, to other cortical points on the same side. Associational neurons may be regarded as interareal, to other functional areas on the same side; and intraareal, to other points in the same area, and these latter may be long or short. There is considerably more interesting detail to the intracortical mechanisms, represented in figure 41 and explained in its legend, which the student is invited to peruse. We might add a few words here about layers ii and vi. Layer ii cells (ff) show a strong tendency to send their dendrites to layer i, thus to derive their stimuli from there. Their axons do not leave the cortex but ramify in supra- and infragranular layers alike. Perhaps they distribute suppression. Layer vi (u-y) cells are smaller than those in most other layers, and give a reduced coverage. Their weak dendrites ascend to outer layers; some axons pass to outer layers (w), others leave the cortex (v) but travel only short distances.

CORTICAL LAMINATION. Differences in dendritic pattern alter forms of cell-bodies, differences in extent of dendritic tree or strength of axon alter their size, and differences in luxuriance of surrounding fibers determine their proximity. Since these factors vary from layer to layer, contrasting pictures are produced which enable the various layers and sublayers to be identified and characterized in sections whose cell bodies only are stained. Thus (figs. 41, 42) layer i is virtually cell free, layer ii is thin but packed with small, dark pyramidal or granular-appearing cells, layer iii is thick and has sparser long pyramids, becoming larger deep in the layer; layer iv is thin and packed with small, pale granular cells; layer v is usually thick, contains pyramids, some of them large; layer vi is generally thick and has scattered, small, fusiform and pyramidal cells. Sublayers are recognized for layers iii, v, and vi. Fiber stains also show differences in layers, with radial fibers from below which diminish at layer iv; and with cross bands formed by clusters of terminals (fig. 41). These have been given special names, but it is simpler and clearer if the bands receive the designation of the cellular layers with which they correspond.

We already have seen evidence that all parts of the cortex are not the same in their connections, and do not perform the same work. Cortex may be classed functionally as sensory receptive, sensory associative, motor, premotor, prefrontal, to name some of the better defined types. From what has already been said in this chapter, it may be expected that the proportions of the various cell types and the relative thicknesses of the layers will differ in these classes, but that they will be fairly uniform within any expanse having similar connections. The cortical pattern types that result are called cortical areas. Brodmann, around 1907, distinguished some 50 areas in the human neocortex and gave them arbitrary numbers. Many years later von Economo, disregarding Brodmann, made a very detailed study, and ended with a map much like Brodmann's, but used letters. The Brodmann designations have become entrenched, and we shall follow them there. I M and VI L show the locations and boundaries of all of them in the human brain except a few which are small in man. Other mammals have comparable areas

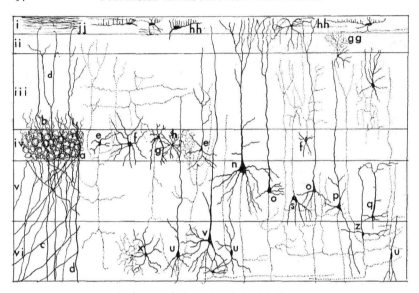

Figure 41. GENERAL STRUCTURAL PLAN OF NEOCORTEX

The greater part of the figure is formed of specimens of neurons in their natural positions, but not from any single areal unit. While a great variety of interconnections is possible, the main circuits follow certain sequences.

The greater number of afferents (a) break up in layer iv among the granule cells. These are generally regarded as thalamocortical, but the development of layer iv in any field does not parallel the intensity of thalamic termination there. Some terminals (b) continue into the basal zone of iii. The thalamocortical fiber stalks (c) are coarse and have a tendency to sweep diagonally through the basal cortical layers. Unspecific and association afferents (d) run vertically and do not form a fine plexus at any one level.

The granule cells of layer iv never send axons out of the cortex of their immediate vicinity. A well developed example (e) sends an axon into the infragranular layers and collaterals into the supragranular lamina. Some ramify exclusively in supragranular layers (f), others only in the infragranular zone (g). The axons of some break up profusely (h) in the immediate vicinity of the cell body, but in the optic cortex the small granule cells have not only short, but sparse, axons (j). Characteristically, they have no terminal dendrites, but some send a dendrite to layer i (k), or have a double bouquet of dendrites (m). This splitting of

the intracortical relay of afferent impulses establishes two basic circuit types: a short, through the infragranular layers only; and a long, utilizing a supragranular-infragranular sequence. A prepotent stimulus can spread from iv directly to a large pyramidal projection neuron in v (n), and so to a primary motor neuron. Such a neuron, however, sends lateral dendrites also into the supragranular layers, and a branched terminal to layer i, so, ordinarily being played on by a wide variety of influences, summates them algebraically. Some pyramids of v show the same dendritic plan, but are smaller (o), hence less inclusive in their collecting power and transmit shorter distances. Some have weaker terminal dendrites (p) dwindling away in iii, or not even leaving v (q). Among these are numerous small, nonpyramidal neurons with short axons (r). The Martinotti cell (s) is reversed, with basally-directed dendrites and an axon terminating in supragranular cortex. All layer v pyramidals have recurrent collaterals (t). Their axons leave the cortical units in which they are located to project to primary motor cells, to motor control cells or to other cortical sites, close or distant.

The neurons of layer vi are more varied, being fusiform (u), triangular (v) stellate (w), or granular (x); more rarely, pyramidal (y). Terminal dendrites, when present, rarely reach layer i (v), the layer vi neuron does not come under strong influence of the thala-

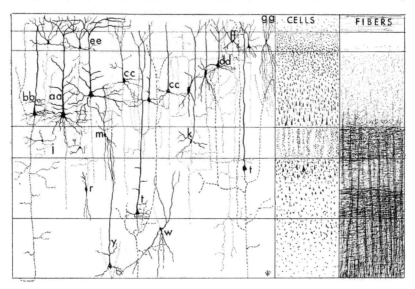

(continued from preceding page)

mic afferents or the supragranular layers; but in general, absorbs the general state of activity in the unit and projects it to nearby units or to the corresponding point in the opposite hemisphere through the callosum. Some (z) merely pass a system of collaterals to layer vi, and to other layers.

The neurons of the supragranular layers ii and iii are preponderantly pyramidal, the smallest in ii, and increasing basally. The typical large pyramid (aa) of basal iii has basal, lateral and terminal dendrites and sends its axon to adjacent cortical points, say, within the same area. They communicate widely within the areal unit by widespread collaterals (bb). Further out, pyramids are smaller (cc), and may or may not show a terminal dendritic arborization. Further out, pyramids are still smaller and show shorter terminal dendrites. Pyramids of ii show a greater reduction, and may lack a terminal dendrite, although they regularly break up in i. The axons of such small pyramids do not leave the cortex, but may

show very fine collateral ramifications (ff). Cells of short axon of various types are numerous (gg).

Layer i is remarkable for its horizontal neurons (hh), which send their axons horizontally within the layer, forming the tangential lamina, of unknown function.

When all the cell bodies are stained, but none of the processes (strip labelled CELLS) a characteristic layering is formed, and the relative numbers of the various forms and sizes impose a pattern. Since the several layers vary in development from place to place in the cerebrum, recognizable type-patterns are formed, termed areas.

When only the fibers are stained (FIBERS) a pattern is similarly formed, but fiber development is largely the reciprocal of cell development.

The exact conditions under which the various cortical elements function are not known; this account is based on deductions from verified observations on structure.

(Combined from Cajal, Lorente de No, Starr)

to a large extent (see figs. 16 and 22 for rat, 24 for cat, and 25 for monkey) and they are numbered similarly, but the total becomes reduced as we pass down the scale.

By closer study it is possible to subdivide some of the areas, even in smaller animals, such as the monkey. In many locations where cortex passes from the gyrus into a sulcal wall, it changes type, and at the bottom of the sulcus (fundus) it changes again. Such types may be within the definition of a single area, although they differ considerably in thickness of layers, number of fibers and proportion of the various connections. In discussing the several areas, we shall take up their appearance in section, and correlate this with the special requirements of the work they do.

CONVOLUTIONAL PATTERN. Naturally, a more active or a more intelligent animal will have a larger expanse of cortex than a sluggish species. This may be expressed in formation of convolutions or folds on the cerebral surface, since the mass of projection fibers to serve it is overbalanced by the wealth of intracortical associations. However, the same relation exists between a large species and a small species of similar intelligence, since the number of cross connections necessary within the cortex increases geometrically as the number of entering neurons increases arithmetically. Thus the cortex of large animals is convoluted, while the cerebrum of small animals is smooth, assuming both have equal intelligence. These two factors conspire to make the human brain highly convoluted, until two-thirds of the cortex is hidden in the folds.*

At first sight the human cortex represents a confused mass of convolutions which could scarcely be traced, named and standardized. As we explore the sulci, or valleys, between the gyri, or ridges, we find certain of them are deeper and longer, and that many of the folds are tributaries. Moreover, if we trace back the convolutional pattern into fetal life, we find it simplifies to a uniform pattern. Among the primates the convolutional pattern is on the same plan. Many of the convolutions show a definite correlation with the functional cortical areas. In fact, to a large extent they are formed in the fetus by the differential growth of areal units consequent on the development of separate systems. Increase of cells and synapses, outgrowth or ingrowth of fibers and multiplication of intraareal fibers expand and stiffen the cortex, so the intervening region must buckle or fold inward. Thus, observed differences in connections between gyral, sulcal wall, and fundic cortex have a sound developmental rationale.

The complex convolutions of the human cortex may be reduced to a single plan, which may be traced in diachrome sheets I, and VI. The deepest of all is the lateral fissure (278) which separates off the temporal lobe (57) (see figs. 52, 53). The next in size is the central sulcus (279), obliquely vertical, separating the cerebrum into two roughly equal halves, the frontal lobe, and the parietal plus occipital lobes, and separating the sensory realm, behind, from the motor realm, in front. On the medial surface (I) are the long cingulate sulcus (62), paralleling the callosum (65), which separates

---

* Man has not the most convoluted cortex, however; the aquatic mammals and the elephant present a veritable labyrinth of convolutions.

the frontal lobe above from the upper arm of the limbic lobe, below. The calcarine sulcus (64) has already been encountered as the axis of the visual cortex. It is in the middle of the occipital lobe. Above it, and joining it anteriorly, is the very deep parieto-occipital fissure (63) separating the lobes indicated by its name. These are the primary fissures or sulci. The constant ones remaining form a simple enough plan. In front of the central sulcus is the precentral gyrus, limited in front by the precentral sulcus (280), sometimes incomplete. Behind the central sulcus is the postcentral gyrus, limited by the postcentral sulcus (284). In the frontal lobe is the massive superior frontal gyrus forming its dorsal shoulder, limited below by the superior frontal sulcus (281). Below this is the middle frontal gyrus, limited below by the inferior frontal sulcus (282), and followed below by the inferior frontal gyrus. Actually, the inferior frontal gyrus consists of three bulky flaps — opercular (44), triangular (45) and orbital (47) parts — with cortex on both surfaces, which, when pulled apart, reveal a large submerged island of cortex, the insula (266: VI M; figs. 52, 53). The parietal lobe is divided into superior and inferior lobules by the irregular intraparietal sulcus (285) and the temporal lobe is divided by the superior (286), middle (287) and inferior temporal sulci into superior (22), middle (21) and inferior (20) temporal gyri (see also figs. 52-54). Beyond the last named, on the lower surface, is the fusiform gyrus (36: figs. 52-54), running from temporal to occipital poles. This is bounded medially by the collateral sulcus (figs. 53, 54). Still beyond, medially is the hippocampal gyrus (28: figs. 52-54), to be distinguished from the hippocampal formation beyond.

INTERNAL ARCHITECTURE OF THE CEREBRUM

The cerebral cortex can be crumbled piecemeal from the gyri and sulci of a cerebral hemisphere which has been hardened in formalin. This is, incidentally, the proper preamble to a fiber dissection of the cerebrum. In the resulting preparation, the gyri stand out and the full depth of the sulci can be seen. The principal convolutions are emphasized, the secondary ones suppressed. It is composed of the fibers that serve the cortex, the alba or white matter, although inside it are the nuclear masses of the thalamus and striatum, and the narrow cavity of the lateral ventricle. Its lateral and medial surfaces fit into VI M and I L, respectively. The gyral cores of the alba consist principally of radial fibers as represented by the outer parts of the fibers in IV. In the sulci is a variable covering of arcuate fibers, which connect adjacent gyri (VI M). These are among the shorter associational fibers of the cerebrum. Others run along the length of the gyri near the surface. Deep to the sulci is the great rounded mass of fibers forming the medullary center of the alba (figs. 52-54). This is definitely organized. Its principal feature is the corona radiata, the converging mass contributed to by the various gyri, recognizable in IV where the fibers first form a continuous wall. Most of it lies in a sagittal plane, but behind, it curves around laterally to send an extension to the temporal lobe (221, 222: IV L). The outer part of the corona is formed of association fibers, the long ones deeper than the short ones (VI M). Medially, the corona sends off the callosum (65: II; figs. 52-54), which carries fibers through the opposite corona to the cortex. Ventrally the corona is continued as the internal capsule (218-220: IV M), a severely compressed lamina containing all cortical connections to and from the thalamus and to (but not from) lower structures. The remainder of the internal capsule continues into the midbrain as the cerebral peduncle (187: IV M). Each point in the cortex has, potentially, at least, its representation in the internal capsule (IV), the thalamic radiation (III), the callosum (II) and in a less regular way among the associational fibers (I L and VI M). A large part of our attention in the study of the cerebrum will be to locate and trace these connections, and they have been worked out in the diachrome series to a degree and with a specificity never before attained.

Also included in the cerebral hemisphere are the massive thalamus and striatum. The thalamus (II) is a rounded mass which sits almost precisely in the center of the cerebral sphere, and the term applied to that region, centrencephalon, by Penfield is well chosen. The thalami of the two sides are separated only by the slit-like third ventricle (71, 72: I). Laterally, the thalamus forms the medial boundary of the internal capsule (218-220: figs. 52-55). Curving around the thalamus is the comma-shaped caudate nucleus (126-128: II). It is likewise medial to the internal capsule (figs. 52-55). The part of the capsule lateral to the main mass of the caudate is called the anterior limb (218: IV); the part lateral to the thalamus is the posterior limb (220). There is a slight difference in angle of the plane of the lateral face of the caudate and that of the thalamus (fig. 55); the apex of that angle is the genu of the internal capsule (219). Lateral to the internal capsule is the lenticular nucleus (223-224: V; fig. 52, 53, 55), forming, with the caudate, the striatum. The internal capsule is sandwiched between lenticular, laterally, and thalamus plus caudate, medially. Proceeding posteriorly, fanwise, the occipitally directed fibers behind the putamen form the retrolenticular internal capsule (221: IV L; fig. 54, 55); while the diminished portion continuing under the lenticular nucleus is the sublenticular portion (222: IV L; fig. 52, 54). Outside the flat outer surface of the lenticular is a much thinner lamina of fibers, the external capsule, composed of long associational fibers (272, 275, 276: VI M; fig. 52, 55). Outside this is a very thin lamina of cells, poorly understood, the claustrum (fig. 52: CL); and between claustrum and the insula is the thin extreme capsule (fig. 52) composed of the fibers serving the insula. The insula is a part of the cerebral cortex which appears to have early become stuck to the lenticular, and the remainder of the cortex has grown in billows around it (fig. 27; p. 115).

The lateral ventricle, originally the hollow cavity of the cerebral vesicle, has become compressed and distorted by the space demands of all the structures detailed in this section (98-100: I L). Its main part, consisting of anterior horn (100) and body (99), is placed between the callosum above, and the caudate nucleus and dorsal surface of the thalamus below. The appearances in section are shown in figs. 52-55, except that the ventricle appears a little distended. In front the two ventricles are separated only by the thin septum (66: I M); behind, they diverge, and curve downward, then forward into the temporal lobes as the inferior horns (97: I; 232: V; figs. 53-55). Along with the inferior horn runs the attenuated tail of the caudate (229: V; 128: figs. 53-55) on the concavity of its curvature. The hippocampus (230, 231: V) is confined in its extent to the inferior horn of the lateral ventricle, but the fornix, which arises from it (67: fig. 54), continues (67: I) in the concavity as far as the interventricular foramen (73: I). Where the inferior horn joins the body of the ventricle, on the convex aspect behind, the lateral ventricle forms a pointed spur into the occipital lobe, the posterior horn (98: I L), so that the lateral ventricle may be regarded as wishbone-shaped.

We now come to the discussion of the anatomy and physiology of the more important cortical areas. Each area and its specific connections are designated by a specific color, and the colors have been distributed to the several areas according to a scheme in which the color indicates something of the function, as a reference to the color key on any opening of the diachrome will indicate. Red is the color representing pure motor, and blue the color for pure sensory. Since there are three main senses involved, somesthetic is purplish blue, visual is light blue, auditory is dark blue. As associational elements come in to each of these, yellow is added, the motor red becoming orange for premotor areas, and the sensory blue becoming green for the parietal areas. The purely psychic or intellectual areas of the frontal pole then become pure yellow. As sensory association cortex acquires the temporal lobe character, the green is browned, proceeding through olive greens to pure brown at the pole. The olfactory system is a different realm, so the purples are used for that system. The olfactory nerve and tract are bright violet, the secondary connections are a mauve, and the hippocampus and fornix are a purplish pink, being efferent. The cingulate areas, not belonging to any of these systems, and not understood, are greyed, but because of their olfactory origin, have a purplish tinge.

With each area we shall want to know its position, special features of its structure, its afferent and efferent connections, its function, and the results of excitation or removal. These topics will be considered in that order, and proportionally to the importance of the area and state of our knowledge about it.

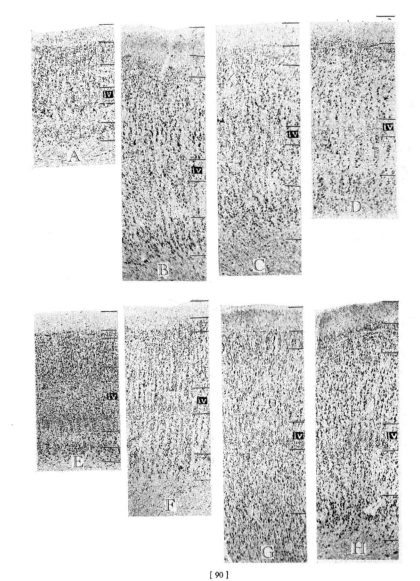

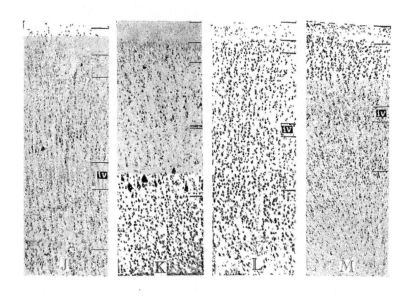

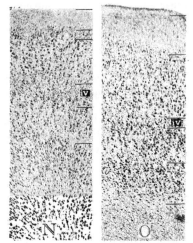

Figure 42. SAMPLE STRIPS OF SOME OF THE PRINCIPAL CORTICAL AREAS IN THE HUMAN BRAIN, STAINED TO SHOW THE CELL BODIES

The six layers are marked off by lines; layer iv is so labelled.

A. Area 3	H. Area 40
B. Area 1	J. Area 22
C. Area 2	K. Area 4
D. Area 7	L. Area 8
E. Area 17	M. Area 9
F. Area 18	N. Area 12
G. Area 41	O. Area 23

(Photography by Miss Madge Walsh)

ANALYSIS AND SYNTHESIS OF SENSATION

SHEET OR NUMERICAL DESIGNATIONS WILL NOT BE CITED IN THE REMAINDER OF THIS CHAPTER AND THROUGHOUT THE NEXT AS ALL LATERAL AREAS ARE ON VI, ALL MEDIAL AREAS ON I, CALLOSAL CONNECTIONS ON II, THALAMIC RADIATIONS ON III, PROJECTIONAL CONNECTIONS ON IV, AND ASSOCIATIONAL CONNECTIONS ON VI M AND I L. *THE STUDENT SHOULD REFER TO THE DIACHROME FOR EACH DETAIL OF THE CORTICAL CONNECTIONS MENTIONED IN THE REMAINDER OF THIS CHAPTER AND THE FOLLOWING CHAPTER.* The chart on the back of the diachrome shows the cortical connections in the form of a hook-up diagram.

ANALYSIS OF BODILY SENSATION. As described on p. 75, the somesthetic system reaches the cortex along the posterior wall of the central sulcus in area 3. The sulcus is deep and area 3 barely reaches the surface. It is depicted in purplish blue, and may be traced in VI L and especially in VI M, where its fundic part is seen. It extends a short way onto the medial surface.

In microscopic section (fig. 42: A) the cortex is of pure sensory receptive type. It is very thin, poorly differentiated but rich in uniform small cells. Layer iv is fairly thick, and there are no large pyramids in layer v or the basal part of iii, as in most other cortex. There is an intense fiber plexus in its middle zone with the characteristic, coarse entering fibers of the thalamus evident. This cortex is adapted to reception, as shown by the thick iv and the terminal plexus; and to emission of only short associations as seen in the small cells of v, but not much to integration, as indicated by poor development of iii. We shall see that by keeping a few principles in mind, an inspection of the architectonics of the various cortical areas will furnish valuable insight of the interpretation of their function.

Its chief afferents are the somesthetic radiations from the somesthetic (VPL) and arcuate (VPM) nuclei of the thalamus. It is singularly free of afferents from neighboring important areas. Area 3 sends fibers to all the somesthetic psychic areas (1, 2, 5, 7, 40). This is the way somesthetic impulses are distributed for processing. Many fibers pass to the motor area 4, and some to the premotor area 6. This enables the most direct sensorimotor connection the cortex can have. Most of these connections, forward and backward, show a tendency to connect with the regions directly across, which means, to the region controlling the same part of the body. Area 3 sends a strong corticothalamic connection back to the same thalamic nucleus. This is an example of the feedback principle whereby weak stimuli are intensified or strong ones suppressed. Very few fibers continue to the peduncle, hence area 3 does not directly control motor nuclei. Its callosal connections are not to area 3 but to the parietal areas behind. Obviously the purpose of area 3 is to receive somesthetic fibers, to distribute them for higher syntheses, and for immediate activity of the nature of a cortical reflex.

In general, the dermatomes (p. 10) of the body are represented in reverse order (VII). Anus and genitals (S3-5) are over the dorsal edge of the medial surface. Then follow, in order, back of lower limb (S1, 2), foot (L4-S1), front of leg (L4, 5), front of thigh (L2-L4), abdomen (T9-L1), chest (T1-T8). This takes us one-third the way down. The cervical segments are represented in the middle third, but in the reverse of the expected order: neck and shoulder are uppermost (C3-5), then follow arm, forearm and hand (C6-T1). The digits are individually represented and occupy considerable space. They are in order with the little finger at the top and the thumb at the bottom of the upper limb zone, which accounts for more than the middle third of the posterior wall of the central sulcus. The lower third is the face zone, with the highest cervical segments at the top, then the trigeminal nerve in the order ophthalmic, maxillary, mandibular. This completes the somesthetic coverage of the body, but just above the lateral fissure, and extending onto its hidden sulcal wall (fig. 52) is a representation of the viscera which has ascended the brainstem by some undefined pathway and has, perhaps, been relayed by the medial segment of the arcuate nucleus, which is of a different cell type from the main or lateral division relaying facial sensation. They continue in the sequence pharynx, larynx, stomach.

Area 1 lies immediately behind area 3, but is on the exposed surface of the postcentral gyrus (light blue). It also hooks onto the medial surface. It is much thicker than area 3 (fig. 42: B). The basal cells of layer iii and some cells in v are medium-sized, making layer iv, between, stand out. The fiber plexus is much reduced, compared to area 3. Area 1 is the only other area besides 3 which receives the somesthetic radiation directly, but it gets only a fraction of the fibers, perhaps collaterals. It also receives fibers from the sensory psychic areas 2 and 5, and from motor area 4, and premotor 6. Area 1 sends off two very strong associational connections to areas 7 and 40 behind, and to area 4 on the other side of the central sulcus. Thus, its purpose must be to distribute sensory impulses to the effector part of the parietal region and to the motor pathway from the cortex. It does not have direct influence on the subcortical motor centers or nuclei as virtually no fibers pass into the midbrain. There is a feedback to the somesthetic nuclei of the thalamus, but it is not nearly as large as the one from 3. In contrast to 3, the callosal element is large, some fibers ending in area 1, some in 3.

In experimentation on somesthetic cortex, it is usually area 1 more than the hidden area 3 that is stimulated, so the functional facts we have are not on the pure somesthetic receptive cortex. If a recording electrode is held over the proper cortical spot, a current will flow when a point on the body is touched. By this method the body-plan was delineated in detail in the chimpanzee. When a point on the postcentral gyrus of conscious human patients is stimulated at operation, tingling is felt in the portion of the skin corresponding. If there is an irritative focus on the postcentral gyrus, the patient will feel a tingling or have an illusion of convulsive spasms in the part corresponding to the region irritated. These are sensory auras. It is in the nature of cortical activity that such irritative reactions spread to adjacent cortex, the sequence of symptoms depending on the connections of the adjacent cortex. Such a sequence is an epileptic attack. Any area of the cerebrum may be the focus of an epileptic attack, and a knowledge of the functions and arrangement of the cortical areas is of the greatest value in localization.

If the postcentral gyrus is removed, tactile localization and higher syntheses are lost on the opposite side of the body, but it cannot be said that anesthesia results. Instead, the affective qualities of sensation-poignancy of pain, unbearable tickle — are exaggerated. From this we can only conclude that such aspects reach consciousness in the thalamus. Indeed, in rare cases when the artery to the posterior thalamus is affected, the thalamic syndrome results, the chief characteristic of which is that slight contacts produce unbearable pain. Apparently the corticothalamic fibers are important in maintaining a proper affective level in our sensory life. It has not been possible to demonstrate that the nociceptive modalities are located in area 3 and tactile localization in area 1, as might be expected. We do know, however, that the higher somesthetic syntheses require functioning of parietal cortex further back.

Area 2 is a strip directly behind 1 on the posterior wall of the postcentral gyrus, that is, in the postcentral sulcus. It is rendered in greenish blue. It (fig. 42: C) resembles area 1, but its cells are closer together and a little more differentiated, especially layer v. Layer iii is thinner and lacks large cells. This was the general type used for the description and illustration (fig. 41) of the mechanism of the cortex, and the best understood.

Area 2 receives its afferents from areas 3 and 1 — somesthetic impulses passed on probably without synthesis beyond tactile localization. It receives few or no fibers directly from the somesthetic nuclei, but probably gets its thalamic supply from the ventralis posterior inferior (VPI) (150: II; fig. 37). The other areas sending fibers to it are the same as the ones it supplies, so these may be the usual reciprocal relations. Judging from its efferents, area 2 is a fundamental unit of parietal organization. It connects with the medial longitudinal fasciculus (182), and it sends numerous fibers down to all three nuclei of the somesthetic group: VPL, VPM, VPI. The efferent fiber connections of area 2 are quite different from those of any other parietal area and are suggestive of its functions. It sends only a few fibers through the peduncle, and these end in the pontile nuclei rather than in motor structures, but a tract leaves the internal capsule, cuts through the posterior part of the thalamus, the lower fiber layer of the superior colliculus, and ends at the rostral end of the MLF, where there are two nuclei (nucleus of Darkschewitsch and interstitial nucleus of Cajal), which send long axons down the MLF to motor nuclei in stem and cord. The callosal connections to the opposite side are heaviest of any area yet studied. They not only end in area 2, but all the surrounding areas 3, 1, 5, and even the motor area, 4. Its ipsilateral connections are equally heavy. The greater part pass to area 4, scattering widely within the area. Many end in 1, also extensively, but none in 3. Others enter the inferior lobule, but there are few to the superior parietal lobule. A component even enters the cingulate cortex. An important link to the frontal lobe is furnished by fibers which run forward in the opercula below the central sulcus to the areas in the inferior frontal convolution (VI M).

The function of area 2, as determined by its connections, is to distribute somesthetic impressions wherever they can be utilized: to more posterior parietal areas as information to work with, to forward parietal areas as feedback regulation, to the opposite side for bilateral synthesis, to the motor cortex for immediate muscular response, to the basal frontal cortex as material for intellectualization, and to the MLF for suppression of motor activity, since, when area 2 is stimulated in an anesthetized animal, a prompt suppression of all motor activity ensues. Results from stimulation are only the algebraic sum of all influences and do not give the complete picture. However, it is possible that the effect of all the multifarious connections of area 2 is inhibition. At any rate, the chief direct influence of area 2 is on the cortex, not on motor or pontile nuclei. It is distributive and integrative, but not psychic or mnemonic.

Areas 3, 1, and 2 are usually considered as a unit, more from ignorance of their differences than any other reason, but they may be regarded as the somesthetic receptive, localizing, analyzing, and distributive areas. They show cortical and reciprocal thalamic connections almost exclusively. They determine motor reactions dependent on sensory recognition, but they do not extensively synthesize the somesthetic modalities, combine other senses with the somesthetic, produce complex images, record memories or intellectualize on bodily sensation. However, they do relay impulses to the areas that perform these syntheses. The postcentral-precentral gyrus connection is important as one of the fundamental connections of neocortex.

Extending forward from the lower end of the postcentral sensory strips, above the lateral fissure, is a cortex of sensory type, the second sensory area, 43 (dark blue). Operative stimulation or epileptic irritation here produces a sensory aura. The body is re-represented, but right side up.

Areas 5 and 7 may be considered together as the superior parietal lobule. Area 5 is just behind area 2, chiefly on the posterior bank of the postcentral sulcus, but on the medial surface it forms a sickle shape under 3, 1, and 2, and touching area 4 above the cingulate sulcus. Area 7 occupies the wide swath between postcentral sulcus and parieto-occipital sulcus on medial and lateral surfaces. It stops below at the intraparietal sulcus, and forms the entirety of the superior parietal lobule.

Area 5, like 2, is found in the postcentral sulcus, and resembles it except that it contains some giant pyramids in v, which project fibers to the pons. Area 7 (fig. 42: D) resembles area 1, but its cells are more widely spaced, and the supragranular layers are thicker, indicating a more integrative function; while the pyramids of iii and v are better developed, evidencing long associational and projectional efferents, respectively.

These areas receive only a few thalamic fibers (156: III); they are from the nucleus lateralis posterior (147: II; figs. 37, 54). Their principal afferents are relayed crude somesthetic stimuli from area 3, and some localized impressions from 1 and 2. Their most striking efferent tract is a strong, cortico-pontile connection, which runs through the capsule and lateral fasciculi of the peduncle into nuclei of the pons (189: figs. 50, 51). Completion of this circuit by the pontocerebellar tract enables sensations of weight and integrated information on the state of muscles and joints to be used to effect a distribution pattern of muscle tonus over the body.

The other important connection of area 7 is to area 4, to incite controllable motor activity dependent on impressions of weight and posture. We have now seen four somesthetic areas acting on area 4: areas 3, 1, 2, and 7, offering somesthetic influences of successive degrees of refinement. All control of the motor area thus does not come through the premotor area 6, much of it is directly parietal. Topical localization must break down in 2, 5 and 7. The superior parietal lobule does not connect with the inferior parietal lobule in monkeys. In man it sends information for higher integration and memory to area 37 and the temporal lobe; and to area 19 for combinations with visual images (see VI M: green). These fibers help to form the extreme sagittal stratum. The highest purely somesthetic syntheses occur in area 7, for if its extirpation is added to that of the postcentral gyrus, in apes trained to discriminate weights and textures, these faculties are lost.

The inferior parietal lobule is chiefly somesthetic, but comes under visual influence, so is best discussed after the occipital lobe is studied.

ANALYSIS OF VISUAL SENSATION. The visual projection system was traced into area 17 on page 71, and the topology of the visual field was summarized there, and shown in Sheet 0. Area 17 and its connections are rendered in the bright blue of the optic system. The cell arrangement of area 17 is quite uniform in the calcarine sulcus or on the gyral surface. This indicates that position on the convolution does not directly influence cortical pattern. The peripheral field has a simpler structure, however, than the binocular field.

The microscopic appearance is one of the most striking and distinctive in the entire neocortex. Area 17 (fig. 42: E) is the thinnest neocortical area, yet the most crowded with cells; its cells are the most uniform, the smallest, and the usual 6-layered pattern the most distorted. At first sight there are three dark granular strata alternating with light bands. In fiber preparations the lower two light bands are dark with fibers, the famous stria of Gennari, discovered with the naked eye in 1776 by a medical student. Professors, always happy to give a student any possible credit, have retained the name. When more closely studied, layer iv is seen to occupy the entire middle third (or more) of the cortex, to contain great numbers of granular cells smaller than a red bood corpuscle, and to be divided in the middle by a light band (of fibers from the lateral geniculate). The outer and inner sublayers have been thought to represent heterolateral and ispilateral eyes because the split occurs in animals with binocular vision. Binocular impressions could then be made to match and to be compared by a system of cross connections. Complex as this may seem, it is the simplest solution of a very difficult problem. Thus, the outer light band is i; the outer thick, dark band is ii, iii and outer iv, as iii has no true pyramidal cells; the middle light and dark bands are in iv; the inner light band is v, thin and poor in cells; the inner thin dark band is vi-a, which is dense-celled; vi-b is thin and pale.

In Golgi sections, axons of iv are seen to distribute to supra- and infra-granular layers; the pyramids of iii lack terminal dendrites and send their axons down to v. Layers v and vi are exceedingly thin, but layer v has pyramids of two types. The small pyramids send their axons into the supra-granular layers. The other type is the giant cell of Monakow, which has extensive, very horizontal basal dendrites, a heavy terminal dendrite to iv and an axon which runs via the internal capsule to the superior colliculus (not illustrated). This is the effector neuron of the optic cortex and mediates the accommodation and convergence reflexes. When objects are viewed at close hand, the eyes must converge, the lens must become more convex to maintain sharp focus, and the pupil contracts to give greater sharpness to the image. All this depends on visual recognition and the necessity of matching the images on the two eyes, so the cortex must be consulted, not the colliculus. Convergence is controlled by the nucleus of Perlia situated be-between the two oculomotor nuclei. Accommodation is effected by contrac-tion of the ciliary muscle, thus permitting the lens to spring back to a thick shape, and is controlled by an accommodation center in the pretectal region, between thalamus and superior colliculus. The effector neurons are in the nucleus of Edinger-Westphal (p. 50).

The main efferents of area 17 are from layer vi-a to the surrounding area 18. Visual signals are identified in position, intensity and perhaps in color in area 17, but this topologic pattern is not transferred bodily to 18. The connections are numerous and densely interwoven.

When part of area 17 is destroyed, blindness results in the corresponding part of the field. In fact, localized occipital injuries incurred in World War I were utilized to determine the pattern of visual projection in man. Punctate stimulation on area 17 at operation results in flashes of light or color localized to the corresponding part of the visual field, but the concealment of most of the visual projection area in the calcarine sulcus and the medial aspect of the cerebrum has prevented acquisition of fuller knowledge. It would seem possible to produce some sort of patterned stimuli interpreted as light, by selective stimulation over a grill of minute electrodes perma-nently implanted on area 17. Such an arrangement might be useful to the blind. Epilepsy with irritative focus on area 17 is ushered in by the sensa-tion of blinding flashes of light.

Area 18, the visual association area, surrounds area 17 on all sides (ren-dered in light blue). The boundary between 17 and 18 in sections is very sharp and is marked by a narrow zone with giant cells in layer iii. Although sharply differentiated from 17, area 18 has a similar general appearance (fig. 42: F) because its cells are small, the layers are poorly differentiated, the cortex is thin and the infragranular layers are narrow. However, layer iii is thick, indicating considerable integrative activity, and has some fair-sized pyramidal cells below, iv is thin and uniformly crowded with minute cells, while the heavy radial fiber bundles cause the cells to be disposed in columns. So far as we know, its afferents are derived only from area 17, but probably it also receives from area 19. Its efferents are to area 19. Area 18 is the region where objects are recognized by their appearance. When it is de-stroyed, individuals, objects or food are not recognized by vision. Stimula-tion produces complex visual hallucinations, whether at operation or as an irritative focus of epilepsy.

Area 19, the visual psychic area, surrounds area 18 and completes the occipital lobe. (It is colored light blue-green in the diachrome.) Its only boundary which coincides with a sulcus is medially, where its forms the occipital wall of the very deep parieto-occipital sulcus (63). In structural type it is transitional between parietal and occipital cortex, as it is in function. It is occipital in type in having rather small cells throughout, and a rather undifferentiated infragranular region, but is parietal in having some pyramidal cells in iii and in being of moderate thickness. It receives and combines recognized visual images from 18; and sends extensive associations forward (273) to be combined with images from the other senses. It also projects the occipitopontile tract (fig. 51) to the pontile nuclei. "This is where optical stimuli are integrated into meaningful images and where reading is organized. When the region is injured, there ensues a visual disorientation or virtual blindness on the corresponding visual field, and, if the lesion is on the left side, a loss of ability to read. When the region is irritated, the patient is the spectator of hallucinations and complicated dreamlike images. An irritative focus in 19 may trigger specific and recurring memory fantasies of highly affective nature which become the incubus of a sequence of images and emotional states, which, as the temporal lobe becomes candescent, are transmuted into subjective states of unreality or detachment of the ego from the bodily image."*

THE LATERALIS GROUP OF NUCLEI OF THE THALAMUS. Because most of the nuclear units of the thalamus are so well correlated in connections with the areal units of the cerebrum, we have not included a separate chapter on the thalamus, but are locating and describing the thalamic nuclei as they are encountered along the systems or serving their cortical areas. Thus the ventralis posterior group was described in connection with the somesthetic system, the medial geniculate with the auditory system, and the lateral geniculate with the optic system. Table B (p. 77) is a summary of the principal features of the several nuclei.

The lateralis group cannot be exactly correlated with specific areas but rather with regions. Further research should precise their connections more closely. Virtually nothing is known concerning the afferent connections to the lateralis group. The lateral nuclei develop late in the mammalian series, the rat (146: fig. 18, 19) has only a small, undifferentiated segment of the thalamic hemisphere representing them, but in man they form a large pillow-like prominence, the pulvinar (which means pillow) (134: III) at the back of the thalamus. Three nuclei bear the name lateralis: lateralis dorsalis, lateralis anterior, lateralis posterior; while the pulvinar, developing from the lateralis posterior deserves a special designation and is even dimly separated into subnuclei. Lateralis doralis is the exception in apparently not having a cortical area. It properly belongs in the next chapter for it receives from the striatum (135: II M; fig. 39). It lies on top of the thalamus behind the an-

---

* Krieg: Functional Neuroanatomy, 2nd Edition, 1953, pp. 386-7. McGraw-Hill Book Co., New York.

terior nucleus as an almond-shaped nucleus (133, pale yellowish green: II M; figs. 38, 39, 53, 54). Lateralis anterior (146, bluish green: II L; figs. 37, 53) is in the middle of the dorsal part of the thalamus. It projects (155, bluish green: III; fig. 53) to the frontal areas 4, 6 and 8, and perhaps to the forward part of the superior parietal lobule. The lateralis posterior (147, yellow: II L; figs. 37, 54), which merges into the pulvinar behind, and not well demarcated, projects (156, pale green; III; IV; fig. 39) to the parietal lobules. These form a part of the intermediate thalamic radiation. The pulvinar forms the entire posterior third of the thalamus and presents a uniform cell type that hardly permits subdivision (134, pale olive: II; figs. 37, 38, 39, 55). Its projection is continuous with and behind the intermediate thalamic radiation, and constitutes the posterior thalamic radiation (158, olive: III, IV, M, L; figs. 39, 55). A part of it curves broadly and continues to the tip of the temporal lobe as the thalamotemporal fasciculus (159: III, IV). The pulvinar radiation contacts areas 37, 39 and 40, occipital area 19, temporal areas 21, 22, 38 — a widespread and diffuse connection.

INTERPRETATION OF SOUND. The auditory system was traced to the auditory cortex (area 41), pages 60 to 65. Experimental stimulation has shown that the tonal scale is cortically represented in order, with the high tones forward and the low tones behind. Thus it might be possible to transmit electrical stimuli induced by sounds to the auditory cortex in deaf persons, grouping a limited range of tones in each electrode. Deaf-mutes cannot use speech because they have never heard speech, so cannot learn to make the correct sounds. Possibly such an appliance could be left in place in such afflicted persons long enough for them to learn to talk.

Area 41, the auditory receptive cortex (41, deep blue: VI L; figs. 54, 55; 42: G), has the structure of sensory cortex. It resembles area 1, being filled with minute cells showing very little differentiation or stratification. Its deep infragranular layers make it the thickest of the sensory receptive areas, and it shows marked radial striation, caused by packets of fibers, passing through iv. Golgi studies show the presence of certain special cells, as the auditory cortex cell which has irregularly wandering dendrites and axons; and the cells of double bouquet whose dendritic plan somewhat resembles a sheaf of grain. The small size of the pyramids in iii-c indicates that the interareal associative fibers do not run far, and the lack of large pyramids in v shows that projectional neurons are absent. Sounds come to consciousness in area 41. The tone and intensity can be identified, but there is no interpretation.

Surrounding area 41, and sometimes cutting it into islands, is area 42, the auditory perceptive area. It covers most of the middle part of the upper surface of the temporal lobe and spills out on the lateral surface of the superior temporal convolution somewhat more than does 41 (42, light Prussian blue: VI L; figs. 54, 55). Its structure bears features resembling area 1: but it has medium-sized pyramids in iii-c and v, indicating a cortex which integrates, and associates to neighboring areas. The infragranular layers are relatively thick but not differentiated from one another. Of its connections we know only that it receives its afferents from area 41, and passes its efferents to the surrounding region of area 22. The pulsating electrical stimuli used for cortical stimulation evoke sensations of humming, ringing, whirring and buzzing. It is here that sounds are recognized and harmonies appreciated.

The further processing of auditory impressions will be considered in connection with the temporal lobe, but we are now in a position to discuss the sensory synthetic areas of the inferior parietal lobule.

SPEECH AND APHASIA. Communication by symbols is a complex and many-sided activity of the cortex. Quite separate faculties are: the understanding of spoken speech (auditory psychic), reading (visual psychic), speaking (branchial motor), writing (cervical motor). In addition, there are the various other symbolisms such as numbers, mathematics, music, drawings, maps and plans. Small wonder then that a wide swath of the cerebral surface seems to have these faculties as its principal function, and if lesions are present, one or other may be lost, causing some type of aphasia. In right-handed individuals all the speech centers are on the left side of the cerebrum. Lesions on the right side do not produce aphasia except in left-handed persons.

Area 40, semantic area. Between the lower part of the postcentral gyrus and the lateral fissure is a convolution which bends around the upcurved end of the lateral fissure, the supramarginal gyrus, covered with area 40. It is colored spring green (VI) as being highly sensory associative. The cortex in this region (fig. 42: H) is thick, with large, widely spaced cells, and shows a well-balanced and fully developed set of layers. Layers ii and iv are differentiated by appearing finely granular, layer iii shows a marked increase in size of pyramidal cells passing deeply, the infragranular layers are thick and v distinguished from vi by showing clear-cut pyramids. These appearances indicate that this cortex has both a considerable internal mechanism and good external connections. It is associative cortex in its purest form.

If area 40 is destroyed on the left side, the highest level of the speech mechanism is lost. The subject cannot organize his thoughts to tell an anecdote or make a description, he misses the point of a joke, he cannot repeat something he has heard or read. Organizing power must come before words can be found and put together, and such patients can talk, read and write and retain any memory and intelligence they had before. This type of aphasia is semantic aphasia. The loss of organizing power is confined to expression. Stimulation at operation makes a patient forget what he was going to say, but he can give brief answers.

Area 22, the syntactic area (olive: VI), forms the outer surface of the superior temporal gyrus. Area 22 is functionally complex. The middle third, surrounding area 42, is the auditory psychic area, where spoken speech is understood. Injury here may interfere with understanding of speech, or it may merely produce distortion of sounds, false sounds, or tinnitus (ringing in the ears). The anterior third is a musical repertoire. Stimulation at operation may cause the subject to hear tunes or songs, often forgotten ones. It is the posterior third, next the semantic area, which is concerned with speech. Stimulation here stops speech in progress of utterance, while injuries produce syntactic aphasia. (See VII.) As its name implies, the patient cannot make grammatically correct sentences. In mild forms, prepositions and connections are left out; or again, only a jumble of relevant words can be evoked. With this is usually a word-deafness or inability to understand heard speech, from involvement of the auditory psychic domain. When the forward part is involved, recognition of melodies or appreciation of music is destroyed.

Area 39, the word area, is in the posterior half of the inferior parietal lobule. The deep superior temporal sulcus usually curves far into this region, and area 39 forms a convolution which folds around it, the angular convolution. It is represented in yellow-green, as representing the most removed sensory association area. It is very like area 40 in section. A subject stimulated here is unable to call by name an object or person presented until the current is shut off. A lesion produces nominal aphasia. Common objects cannot be called by name or words understood. Thoughts or feelings are unaffected but the patient literally does not have the words to express them. More than this, the common symbols, like denominations of coins, numbers, maps, musical scores are meaningless. This is asymbolia. (See VII.)

Of course, specific lesions of these three areas, syntactic, symbolic and nominal, are rarely pure; in fact, some authorities doubt the localizability of the separate stages in formulation of speech. More often there is a more or less complete loss of communicative ability. This is global aphasia. There is another class of aphasias, motor aphasias; these will be considered with the frontal lobe. Aphasic symptoms, because the more crippling and the more dramatic, dominate the symptomatology, but with these there are included the older, subhuman somesthetic and visual synthesis not related to speech, such as concepts and recognition of objects which an animal would need to have.

MEMORY AND ORIENTATION. The temporal lobe is concerned with the highest integration of sensory synthesis, and their abstraction to achieve relationship to environment. In the diachrome the parietal green becomes olive, then umber, and finally pure brown as the cortex becomes more definitely temporal in character.

The temporal lobe comprises (1) the superior temporal gyrus area 22, which has already been discussed; (2) the middle temporal gyrus, area 21, (3) the inferior temporal gyrus, area 20, (4) the occipito-temporal region, area 37; (5) the temporal pole, area 38, and an unexplored strip transitional to olfactory cortex on the basal aspect; area 36 (I M; figs. 52-54).

These areas show the temporal type of cortex (fig. 42: J), which we shall characterize, but not distinguish, except to say that as their color on the map becomes less green and more brown, the temporal pattern is more completely developed. We may take the type as representing the most highly abstracting cortex of the sensory realm. Temporal cortex is very thick but its cells are widely separated, indicating a maximum of internal connections. Layer ii is diffuse and irregular but easily distinguished from iii because the latter has such widely spaced pyramids. A well marked iv is present, but is thinly cellulated. The infragranular layers are very thick but the cells are scattered and not well differentiated, so v blends with vi.

The degree of development of the temporal lobe among the various mammals is dependent on the elaboration of the highest integrative activities within the sensory sphere, so is a rough guide to the psychic level of any species. Because it represents an overgrowth of the cerebral hemisphere behind and below, where the insula is adherent to the putamen, its connections are drawn out, and its fibers to the internal capsule must make a wide circuit to clear this obstruction, thereby forming the sublenticular internal capsule (222: IV L; figs. 53, 54). The auditory radiation (p. 65) forms a large part of this bundle, but alone enjoys a direct linkage, because of the propinquity of the medial geniculate (see 214: fig. 54). The efferent contribution to the internal capsule is the small temporopontile bundle. It takes its place in the cerebral peduncle lateral to motor and parietal contingents (II L fig. 51). The thalamotemporal fasciculus (159: III, IV) is a connection with the pulvinar and sweeps in an arc from the posterolateral aspect of the thalamus. The callosal fibers, lying as they do outside the ventricular system, sweep broadly around the inferior horn close to its wall, as the tapetum (151: II L), and lie medial to the optic radiation. A strong projection to the deep layers of the superior colliculus from area 22 has recently been demonstrated.

A number of well marked long associational bundles are related to the temporal lobe. They are represented on VI: M, colored according to their areas of origin. First is the anterior commissure (277: VI M; fig. 52), a landmark in the midsagittal section (68:I), which is the private callosum for the middle temporal gyrus. In lower forms the anterior commissure is a connection preponderantly between the olfactory bulbs (rat, I L), but this component is virtually absent in man. The uncinate fasciculus (276) is a connection between temporal and frontal lobe, enabling the highest integrations of the temporal lobe to be carried to the intellectualizing frontal lobe, a tract which must be large in philosophers. Above it, but functionally separate, and forming a part of the external sagittal stratum, is the occipitofrontal fasciculus (275), the longest fibers of which carry concepts of optical origin to the frontal lobes and *vice versa*. Along the basal surface is the inferior longitudinal fasciculus (274), permitting occipitotemporal correlations. Other temporal and integrated sensory syntheses, such as speech concepts, reach the inferior frontal gyrus by curving over the insula through the arcuate fasciculus (271). Outside these essentially fore-and-aft associational fasciculi are more vertical ones forming the extreme sagittal stratum (273). Its deeper fibers are more oblique and consist of somesthetic and optic associations running temporalward for higher integration. A surprisingly large number end in area 37.

Stimulations of the extensive middle temporal gyrus (area 21) may evoke complex hallucinations based on visual memories, just as area 22 was seen to be related to auditory memory. Indeed, fiber dissections of man reveal rich connections with area 19, and Marchi studies show the gyrus to be poor in projections.

The functions of the broad inferior temporal gyrus (area 20: VI, fig. 42A) are unknown, but its cortical pattern and the paucity of extrinsic connections of any sort indicate them to be of a synthesizing nature, perhaps integration of personality from experience and memory; and crystallization of character, which, after all, is the quintessence of experience.

When the occipitotemporal area (37) is stimulated at operation, complex hallucinations are evoked, which are often highly charged emotionally. Epileptic episodes from foci here begin in just this way. Stimulations toward the middle of the temporal lobe produce "an impression of unreality, not only of the sensations, but of the entire environment or even of the subject's total image of himself. One of the manifestations of this is that he feels everything he is experiencing has happened before, or that he is viewing it all as another person, or that the whole situation is absurd or ludicrous. Such illusions come very close to the world of dreams, of day-dreaming and free phantasy, usually so completely lost in the normal adult."* If these are con-

---

* Krieg: Functional Neuroanatomy, 2nd Edition, pp. 391-2. McGraw-Hill Book Co., New York.

sidered as caricatures of normal function, we may recognize body image, individuality and continuity of the personality, self in relation to environ- ment, as the integrating function of the temporal lobe proper.

Area 38, at the temporal pole, exhibits a structure of an extreme temporal type. It receives associations from widespread fields, but sends preponder- antly to the frontal lobe, so may furnish a link between experience and con- templation. The EEG shows the temporal pole to be the focus of temporal epilepsy, manifested by momentary automatic motor occupations of an organized but inappropriate sort, in a state of waking unconsciousness, am- nesia and confusion. Here must be the locus of the all-important reality principle. To the extensive panoply of temporal symptoms must be added upper quadrantic visual defects (251: V); and hallunicated odors, from encroachment on the olfactory cortex near the uncus (36: I; 28: fig. 52).

It has recently been shown that the operative removal of areas 35 and 36 (and/or area 28) can lead to loss of memory for recent events.

We have now surveyed all of the cortex behind the central sulcus. The three great information-bearing senses distribute here, one in each of the three lobes. The sensations come to consciousness, are localized, analyzed, combined, remembered, transfigured into symbols which are interpreted, sublimated into an image of the body, personality and relation to the world. Yet in all these hierarchies there is no suggestion of direct action on muscles, no organization of movement with any completeness, no planning for action, no intellectualization on the future, no inhibition of instinctive reactions, except for the reality principle in the temporal lobe. Everything observed is analysis and synthesis, a completion of the pyramid of the three senses to its apex at the temporal pole. Let us see what the frontal lobe can add to this picture.

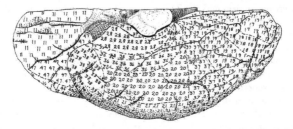

Figure 42A. THE CORTICAL AREAS OF THE BASAL SURFACE OF
THE CEREBRUM

(From Krieg: Functional Neuroanatomy. McGraw-Hill Book Co., New York. Second edition, 1953)

Chapter Seven

CORTICAL CONTROL

EXECUTION

AREA 4, the motor area, with its great efferent tract, the pyramidal system, is the best known of the cerebrum, if not the whole brain. Axons arising from the giant pyramidal cells contact primary neurons of the brain stem and cord all the way down, producing willed movements which are virtually the only final expression of all cortical activity. Area 4 and its efferent connections are indicated in bright red in the diachrome (see also 4: figs. 52-54).

The greater part of area 4 is hidden in the forward or precentral wall of the central sulcus. At the lower end of the central sulcus, area 4 barely reaches the lateral surface (VI L), but the exposed part expands gradually to the dorsal edge of the cerebrum. It extends on the medial surface (I M) as far as the cingulate sulcus, but because the central sulcus does not go beyond the dorsal cerebral edge, all of the area is on the surface.

Area 4 has a distinctive architecture (fig. 42: K). In layer v are scattered the giant pyramidal cells of Betz, by far the largest cells of the cortex, which send a strong dendritic branchwork to all layers, and an axon to the pyramidal tract. There is no layer iv; the pyramidal cells of layers ii and iii gradually increase in size, and beginning with small pyramids at the top of v, enlarge towards its base. Layer vi shows smaller cells which are usually fusiform and not pyramidal. Layer vi melts away gradually into the underlying alba, but even without that, area 4 is the thickest in the cerebrum. This cortex is of the pure motor type.

Area 4 receives afferents from a wide variety of cortical areas. From the sensory sphere the more important have been mentioned: 3, 1, 2, 5, 7. We do not at present know whether other sensory areas than somesthetic connect directly with 4. The motor area is under the closest direct control from the premotor area, 6. From further forward — 8, 9 — the direct connection decreases rapidly, although a chain series 9-8-6-4, etc., is clear enough. Thalamic connections are from nucleus ventralis ventralis (130: II, III), furnishing a strong cerebellar feed-back via the brachium conjunctivum (p. 137); and from VPI (150) and LA (146), as recently shown.

Before tracing its projection system the other efferent connections of area 4 should be mentioned. The callosal connection is very weak, as the two sides of the body are under separate volitional control. Between the opposite face areas, it is somewhat stronger, and the two sides of the face usually act together. Liberal connections are sent to the postcentral gyrus and fewer to gyri across the postcentral sulcus. A moderate number feed back to premotor area 6 in its posterior part, but not to any area in front, or to distant regions. The interareal connections both from and to 4 are largely confined within leg, arm or face zones. Local excitation by strychnine has shown there is more tendency for excitation to spread over the entire span 6-4-1-2-5 than to zones of the same area supplying other bodily members. Functionally, it is useful to think in terms of leg, arm and head paracentral sectors.

Localization of muscles on area 4 goes much further than the main bodily members, even with the crude method of electrical stimulation (Sheets O, VII). In the hand, for example, even individual muscles respond. The plan is parallel to that on the postcentral gyrus, so far as motor localization can parallel sensory. On the medial surface, nearest the cingulate sulcus, in tailed monkeys is the tail region, but in man the sphincters and perineal muscles come first (S3-S5). The buttock muscles, hamstrings and muscles in the foot (S2) succeed these dorsally. The fancied hemi-homunculus hangs by his ankle from the dorsal edge. On the upper third of the precentral gyrus follow in order foot, leg, thigh abductors, knee extensors, thigh flexors. The toes have a wider representation than the thigh. Abdomen and thorax (T1-L2) occupy very little space since there is so little volitional nicety in their movements. The upper limb segments (C5-T1) are reversed, as in the sensory area, so we witness the order, from above: shoulder, arm, forearm, hand, fingers—each successive segment taking a wider swath. In fact, each digit is separately represented, the fifth above, the thumb below. Cortically controlled finger movements are the greatest single motor factor in man's dominance. The head area follows the order: neck, eyelids, lips, tongue, jaw, palate, larynx, as in the somesthetic sequence. Eye movements are not represented here.

The diachrome (IV) shows the arrangement of fibers forming the pyramidal system in medial and lateral aspects, and figures 52-54 portray them in frontal slices. The highest stations in the cortex send fibers vertically (fig. 54), the succeeding ones more obliquely (fig. 53), until the lowest must turn upward in the subcentral operculum to clear the insula and putamen (fig. 52). Reaching the internal capsule the order is face, arm, leg, from front to back within the capsule's posterior limb. They skirt past the thalamus without substantial loss and enter the peduncle of the midbrain in force (fig. 51). Here they occupy the second and third fifths, from medially. In the pons the peduncle breaks up into fascicles but area 4 fibers occupy several large central bundles (fig. 50). At the lower end of the pons, virtually the only fibers left are pyramidal fibers from area 4. They occupy a round bundle on the ventral surface next the median line (fig. 49), and continue in this position through most of the bulb (fig. 48). The fibers are now completely scrambled. At the lower end of the bulb a massive crossing occurs, the decussation of the pyramids (196: IV; figs. 28, 32, 45, 48). Coarse fascicles from the two sides interdigitate, and the crossed bundles resulting are the lateral corticospinal tracts we studied in the cord (p. 16) (197; see also figs. 5, 7). About a tenth of the pyramidal fibers fail to decussate. They continue in the ventromedial part of the cord as the ventral corticospinal tract (198). The tract gradually diminishes by losing fibers to the motor neurons of the cord. The lateral tract fibers do not cross again; the ventral ones cross in the ventral commissure of the cord along the way, at the level of their terminations. A large part of the corticospinal tracts are lost to the lower brachial plexus region (C7-T1) controlling the fingers.

The main pyramidal tract as described takes care of the spinal nerves, but what about the cranial motor nerves? Muscles of expression, tongue and larynx are under the fullest volitional control. Along the way in the brain stem the peduncle gives off groups of fibers to motor nuclei, the corticobulbar tracts; some of these bundles come from the pyramidal system. The confirmed pyramidal ones are detailed below. (Other corticobulbars are described with area 6.)

a) Pontile corticobulbars (190: IV) leave the pyramidal fascicles in the pons and filter dorsally through the pontile nuclei; then, continuing in the midplane, they reach the fourth ventricle and run outward near the facial nerve root (and often are confused with it) to reach the facial and masticator nuclei, chiefly the former. Others of this group cut laterally through the tegmentum opposite the facial nucleus.

b) Pontobulbar corticobulbars (191: IV) bridge the gap between pyramid and lemniscus at the pontobulbar junction; crossing and turning into the reticular substance, they reach the facial nucleus. Some are uncrossed.

c) Interolivary corticobulbars (192) have a corresponding course. They probably carry the organized impulses of speech to the ambiguus nucleus.

d) Direct reticular corticobulbars (193) may be from area 6 to the inferior reticular nucleus, or from area 4 to the hypoglossal.

e) Hypoglossal corticobulbars (194) leave the crossing pyramids and pass directly to hypoglossal nuclei of both sides.

f) Pick's bundle (195) turns rostrally from the decussation and runs medial to the spinal V nucleus. It cannot be traced to any special nucleus but may go to the ambiguus.

Relatively to the amount of detail these are less important than the main pyramidal tracts, but a tegmental lesion may destroy the crossing fibers, producing a pseudobulbar paralysis of cranial motor nuclei, characterized by severe disturbance of trituration of food and of articulation of speech, without disturbance of cortical control of spinal nerves. When the pyramidal tract is damaged in midbrain or pons, the lesion is usually unilateral, and since there is much bilateral control of face muscles from a single side, the cranial nerves, particularly the facial, are not so severely stricken. Lesions affecting the capsular part of the pyramidal tract are the most common brain lesions. The relatively less frequent ones of the brain stem might be localized with the knowledge at the reader's present command. Ordinarily the emergent fibers of one or other cranial nerve will be caught in the lesion. The result is a flaccid paralysis of the head muscles on the same side as the lesion, and a loss of cortical control of the spinal nerves on the opposite side. The cranial nerve or nerves paralyzed enable the level of the lesion to be stated. This is an alternating hemiplegia. Clinical lesions affecting the pyramidal tract, while still in the cerebrum, require knowledge of the premotor area for understanding (p. 111).

When a point on the motor area is stimulated for some time, the excitation spreads to neighboring points, and the discrete contraction spreads until a generalized convulsion of one side results. An irritative focus can do the same in the intact brain and the result is epilepsy of the Jacksonian or motor type. The site of first activity is the key to the position of the lesion.

ORGANIZATION OF ACTIVITY

In front of the motor area lies the extensive premotor area 6, where movement patterns are organized (red-orange in the diachrome). Though all parts appear to belong to one structural type, it is actually composed of a half dozen or more subareas, differing subtly in cell architecture. Like 4, it is broad at the dorsal edge and tapers as it passes ventrally. Above, it lies on posterior parts of superior and medial frontal gyri, below it is limited by the precentral sulcus. On the medial surface it is as broad as at the dorsal surface and stops at the cingulate sulcus.

In cell-stained sections it is much like area 4, but the giant pyramids are absent, and there is a general falling off of the strongly drawn features of 4: thickness, size of cells, richness of fibers. The usual names for one version of the subdividing are rather inconvenient. The broad strip in front of area 4 is 6a alpha, except for the lowest part, which is 6b. The forward extension dorsally is 6a beta. The last named has a thin layer iv and drops off still more from the motor type.

Area 6 receives highly refined direction from the more forward frontal areas we have not yet discussed, and from more forward stations in area 6. As already shown, the areas 3, 1, and 7 furnish a direct somesthetic influence. Its thalamic afferents are from VV(130), derived from the cerebellum; from VPI(150), and LA(146).

Its efferents are numerous and varied. Its projection tract is large and comes mostly from the posterior part. The stalks have the same convergent arrangement as do those of area 4 (see also fig. 52). Reaching the internal capsule, they run directly downward in front of the pyramidal tract, but unlike 4, they give off fibers in the subthalamic region which regulate stereotyped movements. The first of these (182) cuts through the ventral part of the thalamus, and ends in the nuclei at the rostral end of the M. L. F. like a similar bundle from area 2. Other fibers (183) enter the posterior part of the hypothalamus and basal tegmentum. A large and important group is the lemniscal corticobulbar fasciculus (184: IV; figs. 51, 50). They diverge from the peduncle just above the pons and turn dorsally to the territory of the medial lemniscus and down that tract in a reverse direction, the reverse of that taken by its proper fibers. In the pons the lemniscus lies on top of the pontile formation, quite separated from the other descending bundles. In the middle of the pons a group (185) diverges to continue in the upper part of the tegmentum and ends in the reticular nuclei there, indicating it to be an important tract in tonic motor control. The lemniscal group continues (186) into the bulb and ends in the bulbar reticular nuclei. It is likely the upper group is suppressor in effect, the lower facilitatory.

The main mass of projection fibers keeps its position in the peduncle, medial to the pyramidal fibers. There is controversy whether any fibers end in the nigra (163: III) but many pass through it. The great majority end in the pons, among the medial and dorsal cells of this great mass. This is the main medium of keeping the cerebellum in touch with voluntary movements about to be and being executed. The organization plan can be passed on to the cerebellum which can "set" the muscle tonus over the body for the movement. A small number accompany the pyramidal fibers in the bulb; most are exhausted in the reticular nuclei, and very few reach the lateral corticospinal tract. This is probably the only area other than 4 that sends fibers into the cord, although 5 and 8 might possibly.

The callosal contingent from 6 is quite large, in contrast to 4; however, nearly all end in area 6 opposite, not in other areas. Organization of movements needs close cooperation between the two sides. Thalamic connections with ventralis ventralis from 6 (with 130) are more extensive than reciprocally, and probably other nuclei influence area 6. The strongest extraareal associational connections are to area 4, as expected, for the organized acts are passed on to 4 for effectuation. Marchi studies on monkeys have shown that it is chiefly the crests of the convolutions which send out and receive associations. This is a general, but not universal, rule. The subdivisions of area 6 associate strongly with each other. There are several links or stages between the forward end of area 6 and area 4.

The hierarchy of zones in 6 is shown by the results of excision or stimulation. If its posterior part is excised, many skilled movements are lost; the same result can be attained by separating 6 from 4. When stimulated, co-ordinated movements result, confined to one part. If its anterior part is excised, no effect on motion is seen, but if stimulated, massive movements are witnessed, as general adversive movements of the eyes, head and body. Epilepsy of this region is recognized by the violent rotating movements toward the opposite side. These results are not effected by area 4, but by descending fibers. A laboratory test for excision of area 6 is the forced grasping when an object is placed on the palm. This is seen in man by observing the flexion caused by flicking the digits. A corresponding test for area 4 damage is the famous sign of Babinski: extension of the big toe with fanning of the toes on stroking the sole.

A "stroke" or hemiplegia is a massive damage to the internal capsule caused by bursting of a small artery in the basal part of the brain. The result is a paralysis of voluntary movement of the opposite side. With this is a spasticity or stiffness and unnatural posturing of the members. The paralysis is from damage to area 4 fibers, the spasticity from damage to area 6 fibers. Voluntary control of skilled movements never recovers, but other tracts learn to function vicariously for the cruder movements. The spasticity diminishes greatly in time. Epilepsies of 6 spread rapidly to 4, and vice versa; both result in widespread muscular contraction, so are not easily distinguished.

Area 8, the area of adversive movements, is a vertical strip in front of area 6. It is here rendered in middle orange, because it has more associational mixture with its motor nature. Like 6, it is large at the dorsal edge, and below is narrow and drops into a sulcus; medially it echoes 6, as does 4. In section (fig. 42: L) it is of the frontal type. The cortex in front of area 4 shows a gradual change, proceeding 6-8-9-10-11-12. The features are these: thinning of the cortex, general reduction of cell size and loss of pyramidal outline, specific reduction of cell size in iiic and v, appearance and thickening of iv, sharpening of lower edge and narrowing of vi. Area 8 is characterized by resemblance to area 6 but having a narrower granular iv. The features indicate an increase of specific afferents, a decrease of projectional fibers, and a tendency toward integrative association.

The connections of area 8 are not well understood, although details are known, and the long fibers of 8 and all areas more frontal are fine of caliber and sparse. Such cortical afferents as have been seen come from area 9, continuing the general polar to central concatenation already noted. It receives numerous fibers from the front half of the thalamus, from ventralis anterior or from medialis. The callosal connection is not strong, and no capsular fibers are seen below the thalamus in monkeys; but we have functional reason to believe a definite contingent is present in man which leaves the deep surface of the cerebral peduncle as the lateral corticobulbar tract (fig. 51) and cuts through the tegmentum to end in contact with the eye muscle nuclei. This tract is believed to stem from area 8 because when the middle half of that region is stimulated, the eyes are forcibly turned to the opposite side (conjugate deviation), and if continued, the head and body turn. If area 8 is extirpated, the eyes are driven by the intact side, indicating a normal balance of opposing tonus in the normal status. In anesthetized animals, but not in man, stimulation of area 8 produces suppression of movement. The cortical areas which have this suppressive effect are 8, 2, 19, 24 and a strip on the boundary between 6 and 4. The common connection shown by these suppressor strips is to the cingulum (see p. 116).

PRODUCTION OF SPEECH. In the inferior frontal gyrus are the two closely related areas 44 and 45. They are colored a brownish orange to indicate they are a modification of the premotor area. In cell arrangement they resemble area 6 in that ii, iii and v are definitely pyramidal, area 8 in having a thin iv, while the basal stratum of iii has large pyramidal cells. Afferents come over the arcuate fasciculus (271) from the sensory speech areas. The thalamic connections are either with ventralis anterior or medialis, but are weak. In monkeys a projection system is lacking, but there is a strong callosal system, which a unilateral speech area would need in order to stimulate the opposite motor area. Efferent connections to areas 6 and 4 would seem to be essential; however, they seem poorly developed in monkeys (no data for man).

Stimulation here at operation causes an arrest of speech, extirpation destroys the power of oral speech (motor aphasia), but not of speech formulation or writing. Area 44 has long been known as Broca's motor speech area. The writing center is above area 44 at the front of the hand premotor area (VII).

PLANNING OF ACTION. ABSTRACTING

Areas 4, 6, 8, and 44 respond to stimulation and have to do with organization and execution of willed motor acts; the areas forward to the frontal pole — 9, 10, 11, 12, 46 — conceive and will the motor acts, they deliberate on sensory data, they intellectualize and rationalize, they "create." These are the prefrontal areas. Always the highest and the most valuable cortical functions have been regarded as those whose origin is most remote from sensory or motor spheres. However, since when routed through the prefrontal cortex the chain connecting the two end points of reality, sensory and motor, is longer than when perception leads directly to action, there is more danger of a slip or a break. Quickness of sensory apperception and response should not be confused with judgment and planning ability; they have entirely different neural connections. Most of our educational and vocational activities are concerned with the shorter circuits, but effective conduct of life employs the frontal concatenation.

Area 9 occupies the prominence of the forehead. It is on the medial and dorsolateral surfaces of the superior frontal gyrus, but numerous secondary convolutions are found here. Whether any relation exists in man between the number of prefrontal convolutions and the degree of intelligence has never been satisfactorily demonstrated, but it is a fact that there is a great differential between the development of the prefrontal region in man and the anthropoid apes. Area 46 is really a subdivision of 9 and occupies a large, irregular bight in it. Area 10 occupies the region of the frontal pole, extending to the medial surface. Areas 11, 12 and 47 are related to the lower or orbital surface and will be considered separately. Yellows are used for the prefrontal areas, beginning with yellow-orange for 9, orange-yellow for 10, chrome yellow for 11, lemon yellow for 12.

The lamination patterns of areas 9, 10 and 46 show the full development of the prefrontal type (fig. 42: M). The finely granular, thick, layer iv is near the middle, while above and below it are equally broad bands; the supragranular with widely spaced pyramids, the infragranular with a few pyramids in v and a thick, fusiform-celled vi. Fiber stains show very few myelinated fibers of any sort, except that in area 10 there are coarse, horizontal fibers in vi, which gives that layer a characteristic quadrilled appearance.

Area 10 receives long associational fibers from the parietal lobe via the arcuate fasciculus (271: VI M), from the temporal lobe via the uncinate fasciculus (276), and from the occipital lobe via the occipitofrontal fasciculus (275). These furnish sense data for reflection, commentary, integrated impressions and creative intellectualization. This is one of the strong sites of discharge of the medial nucleus of the thalamus (131: II M; figs. 38, 53), through the anterior thalamic radiation (152: III; figs. 39, 52), which joins the forward-running fibers of the anterior limb of the internal capsule (218: IV). The efferent connections of area 10 are to the medial nucleus of the thalamus, through its anterior pole (154: III; figs. 52, 53, 55) as a feedback to the nucleus which stimulates it; to the temporal lobe, intellectually influencing the integrating faculties; and to area 9, furnishing its most important source of intellectualization of experience.

Area 9 receives its data from (1) the medial nucleus via the anterior thalamic radiation, particularly that part forming a compact forward-running funiculus between callosum and caudate* (153: III; figs. 39, 52, 53); (2) area 10, as indicated above; (3) extensive connections from frontal areas of lower order (8, 6, 44). Area 9 and area 46 probably house the highest intellectual functions of creative thinking, planning, judgment, wisdom. The efferents pass to the more effector cortex, areas 8 and 6, where the results of thought are embodied in activity or repression — equally important in the well-ordered life — and to the medial nucleus. Area 46 is apparently a grade more remote even than 9, there being few connections of any kind. All these areas have numerous intrinsic connections and show a fairly good callosal connection. They may send down corticopontile fibers (187: IV; fig. 51) in man (but not in monkey).

The areas 11, 12 and 47 are of the frontal type, but lie on its horizontal orbital surface, and the adjacent medial surface. They are colored yellow in the diachrome to indicate their remoteness from efferent or afferent connections.

Area 11 is of the prefrontal type, but layer iv is inconspicuous, and its thick infragranular layers relate it to the temporal cortex. It does, however, receive many fibers from the medial nucleus. Its most conspicuous associational connection is with the temporal polar region via the uncinate fasciculus (276), carrying information on the integrated bodily state, and is a two-way connection. A strong connection from the insula is hypothecated from knowledge of functions of both regions. Its efferent connections are to the neighboring areas 10 and 12, to which it probably carries integrated emotional states; and to the hypothalamus (p. 147). An experimentation on area 11 indicates it to be the cortical center for emotional states, particularly those connected with well-being, as stimulation here produces peristalsis, vasodilatation, and lowering of blood pressure. (See fig. 42A, p. 105).

The cell-pattern of area 12, medial to area 11 (fig. 42: N) is of the extreme frontal type, being packed with small cells on either side of layer iv, and contains almost no myelinated fibers. The medial thalamic nucleus supplies fibers. Its cortical afferents are from the uncinate fasciculus and from area 11. No myelinated extraareal efferents have been seen, but unmyelinated fibers probably pass to the hypothalamus. Stimulation causes a stereotyped rage response mediated through the hypothalamus. At the risk of oversimplification, we may regard area 12 as the cortical center of sympathetic responses, and area 11 as controlling parasympathetic reactions (see p. 141). Perhaps its function is farther-reaching than the crude responses observed, and have to do with the emotional ordering of the individual. Whether

---

* Not the subcallosal fasciculus, which is more medial, smaller and poorly myelinated.

fibers pass into the peduncle destined for the pontile nuclei, from any of the prefrontal or orbital areas in man is uncertain, but in the monkey none are apparent.

The known relation of the prefrontal region to the psychic and emotional life of the individual has led to experimental frontal lobotomies, not only in higher primates, but also in man, here as an attempt to relieve crippling neurotic states characterized by withdrawal, listlessness and obsessions. Although the methods widely in use today utilize crude and uncritical approaches, a fair number of afflicted patients are relieved of their chief symptoms. Frontal lobotomy reduces the ability and the desire to cope with problems or to inhibit spontaneous behavior. The operated subject becomes indifferent, careless of appearance and speech, irresponsible in behavior, easygoing, boastful and easily antagonized. Other "psychosurgery" has been attempted. A cut into the medial part of the frontal lobe reduces the appreciation of visceral pain; extirpation of the postcentral gyrus diminishes somatic pain; and deracination of an epileptic focus abolishes attacks. Doubtless, with increase of functional anatomical knowledge, new excisions and stimulations will be developed.

The localization of function in the prefrontal region on sheet VII of the diachrome is more schematic than present knowledge warrants.

OTHER CORTICAL REGIONS

The regions already discussed — parietal, occipital, temporal and frontal — constitute true units, whether regarded by their topography, architecture or connections. They are best known because the most accessible; and, it happens, the most significant. There are other regions of the cortex less accessible and less understood. They comprise insula, cingulate region and olfactory region.

The INSULA is a sheet of cortex hidden by the folds of cortex from the orbital, precentral, central and parietal regions above the lateral fissure, and by the superior temporal gyrus below it, but attached to the inner edges of these cortical folds at its boundary, marked by the circular sulcus. (See 266: VI M, fig. 27 and figs. 52-55 for relations.)

Its surface faces laterally and is marked by several shallow sulci radiating upward and backward. Anteroventrally it merges with the orbital and temporal regions; so area 16 (16: VI M) is brownish, but areas 15 and 16 also show olfactory characters, hence are purplish. Its greater part is divisible by an oblique line echoing the central sulcus into a posterior part, area 13, with a parietal type of pattern; and an anterior part, area 14, of a general frontal type. Their tints in the diachrome are paler versions of the parietal and frontal types, respectively. Because it is hidden and because the middle cerebral artery branches freely over its surface, experimental investigation has been backward, but it is now known that the insula is the seat of visceral sensation and motor control, in areas 13 and 14, respectively. Considering the sequence of bodily representation outlined for somesthetic and motor cortices, this is a logical position, the face and mouth centers being located at the lowest part of the free surface, the branchial centers continuing down their series on the under surface of the opercular flaps of cortex, with finally the thoracic and abdominal viscera on the insula. Human stimulations give gastric or intestinal sensations, heart palpitations, or induce gastro-intestinal motility and quickening of heart rate. Stimulation in the region of the circular sulcus around the insular margin gives illusions of taste. The thalamic relay nucleus is presumed to be the medial division of the arcuate nucleus, in reality a separate nucleus (not distinguished in illustrations), with much smaller cells, and few myelinated fibers. The course of the thalamocortical gustatory fibers has not been traced, nor has the secondary gustatory tract. The fibers serving the insula occupy the extreme capsule and do not cross the claustrum (figs. 52, 53).

CINGULATE REGION. The cortex which forms the edge of the cerebral hemisphere all around (25, 33: I M; 230, 231: V M; also see figs. 52-54) remains olfactory in all vertebrates, and the ring of cortex surrounding it is transitional to neocortex. It is sometimes called limbic lobe or gyrus fornicatus. Its lower limb contains areas 35 and 36 (figs. 42A, 52-54). We shall refer to the upper part of it as the cingulate region or gyrus. Its inner edge is in contact with the callosum, its outer boundary is the cinguate sulcus (62) through most of its extent. Its magnitude in lower mammals can be seen in the reconstruction of the rat brain (rat, 24, 23, 29: I M). The cingulate cortex is served by its private association bundle, the cingulum (96: I L).

Although several areas are included, cingulate cortex is a distinct and aberrant type (fig. 42: O). In its purer forms, the cellular layers are uniform and undifferentiated, being packed with oval cells; numerous radial fibers run out to layer ii, which is unique in the cortex. Layer i is very thick and contains a great many tangential fibers. These characteristics are best seen in lower forms, in man a layer iv appears. The cingulate character is most marked in the retrosplenial area 29, small in man, but large in the rat (fig. 16). Extending forward, areas 23 and 24 represent successive reductions of the type, while outside these are the areas 31 and 32, whose structure is transitional to parietal and frontal cortex, respectively, which surround them. All are shown in I M. The cingulate areas proper have a ground coloration of gray in the diachrome to indicate their different, but unknown, function, but with an admixture of purple to acknowledge their olfactory connections; pinkish in front (24) because of their relation to motor cortex, bluish behind (23) to show their relation to sensory cortex. Their outer zones form separate areas which are transitional to neighboring cortex: area 32, merging with frontal cortex, has a further addition of red; area 31, merging with parietal cortex, is tinged with blue.

The cingulate areas receive the radiations from the anterior thalamic nucleus. This major division of the thalamus forms a bun-like prominence on the top of the front end of the thalamus (111: I L; figs. 37-39, 52, 53, 55). The thalamocortical fibers (112) leave by the anterior thalamic radiation and join the cingulum (96). Many of the somatic cortical areas send contributions to the cingulate cortex, particularly the so-called suppressor strips (8, 6, 2). Most fibers entering the cingulum (96) run backward and tend to end far back. Area 29 receives olfactory associations, and it is the source of the heavy tangential fiber lamina of the cingulate cortex. The cingulate cortex of higher forms does not seem to send any fibers into the coronacapsule-peduncle system, and no outlet is known for its impulses. Its function is unknown, but stimulation produces some autonomic effects and suppresses motor activity.

OLFACTORY CORTEX AND PROJECTION PATHWAYS. Reflecting the primary division of the neocortex into sensory and motor divisions, the olfactory areas can be divided into a sensory division which includes primary (area 51), and secondary (area 28) olfactory cortices; and a motor division, the hippocampal formation (colored pinkish purple, as being olfactory, twice removed). The afferent connections and structure of both these divisions have already been described on page 80. The cells of the hippocampus proper may be regarded as corresponding to the Betz cells of the motor cortex, and the fornix as the efferent tract. The fornix arches upward and medially along the inner curve of the lateral ventricle (67: I; fig. 52-55), then forward under the callosum and septum (67: I M; fig. 53), then (67: I L; fig. 52) behind the anterior commissure, through the hypothalamus, and ends in the mammillary body (109: I L; figs. 39, 52). The mammillary bodies (Frontispiece) form a pair of rounded prominences at the back of the hypothalamus, between the cerebral peduncles. Their cells send their axons directly dorsally to the anterior nucleus (111: I L; figs. 39, 52, 53) as the mammillothalamic tract (110: I L; fig. 39, 52). As already stated, the anterior nucleus discharges into the cinguate cortex. The mammillothalamic tract sends a branch into the tegmentum of the brain stem. This is not easily discerned and is overshadowed in prominence by another system, as follows.

In the hypothalamus, preoptic region and olfactory tubercle, a small bundle forms, which runs backward along the medial edge of the thalamus. This is the stria medullaris (107: I L; figs. 39, 52). Most of the fibers end in the small habenular nucleus of the thalamus (113: I L; figs. 38, 39, 54), whose axons pass in a small but conspicuous tract, the habenulopeduncular tract (114: I L; fig. 39) through tegmentum and red nucleus to the inter-peduncular nucleus (115: I L; fig. 39) behind the mammillary bodies and between the thighs of the cerebral peduncles (frontispiece). From here, in turn, a tract descends and diffuses into the tegmentum, placing the reticular substance of the brain stem under olfactory influence.

Placed and shaped like a football, just in front of the toes of hippocampus is a nuclear mass known as the amygdala (228: V; fig. 52). Its connections relate it to the olfactory system, its origin relates it to the striatal complex (whence its brown coloration in the diachrome), but its function in mammals is most unclear. It contains several distinct nuclei. One well-marked tract emerges from its upper surface, the stria terminalis (108: V M, I L; figs. 52-54) and follows a curvature similar to that of the fornix, but marks the dorsolateral edge of the thalamus. It ends in the hypothalamus.

Injury to the amygdaloid region in animals results in emotional changes and hypersexual behavior of an extreme and bizarre nature, but more critical studies indicate that the responsible lesions are in the pyriform cortex, area 28, just below; while stimuli applied to the amygdala proper produce oral activity, as chewing, licking the lips, and the like, suggesting that it is an ancient center for control of feeding reflexes.

The septal region, formerly considered negligible functionally, has come to the forefront as a "pleasure center," because animals with electrodes implanted in the septum, permitted to press a key delivering a stimulus will continue to do so until they drop from fatigue! It receives afferents from the primary olfactory and discharges to the hippocampus over the callosum.

All of the structures considered in this section are most ancient; their counterparts dominate the salamander's brain (figs. 11-13), and they are prominent in the rat's brain, where they may be more easily followed (see corresponding numbers in rat reconstruction). In striking contrast to their anatomical prominence, their exact role is not understood except that they are basically olfactory. This entire system, together with the primary olfactory structures, are grouped together as the rhinencephalon (nose-brain), a structural and phylogenetic unit, rather than a functional one.

It must now be apparent that the cerebral cortex exceeds in importance, significance, interest, and manifoldness of function any other part of the nervous system. Conducing to its satisfactoriness as an object of exact study are the correspondence of separate functions with the recognized cortical areas, the possibility of determining the connections which instrument these functions, the correlation of microscopic structure with functional type, the relatively exposed situation of most of the important areas, and their two-dimensional display. Militating against its investigation are its over-whelming size when surveys by microscopic section are attempted, the large extent of its silent areas, the diffuseness and equivocality of many of its connections and the subjective nature of the function of many of its parts, requiring human material. Considering all these factors, the paradox may be appreciated that the most important and fascinating part of the brain is the least studied and most poorly understood.

The cortex is important in the practice of general medicine, too; but particularly to the specialties of neurology, psychiatry and psychology. Because of its size and its exposed position, there is a greater proportion of cases of localized injury, hemorrhage, tumor and localized infection than in all the remainder of the brain together; while the important condition of epilepsy is confined to the cerebrum alone. A knowledge of the functional localization and relative positions of the areal units, and the arrangement of their connections and projections, founded on a clear visual picture and understanding, is the best equipment for diagnosis, prognosis and treatment of afflictions of the cerebrum. The distinction between an injury resulting in loss of a higher faculty and a deficit in intelligence, or functional mental aberration; and, moreover, between organic damage and hysterical (unconscious) and malingering (conscious) simulation of symptoms may depend on the soundness of the physician's grasp of material in these two chapters — and this is the job of any physician, if only to be able to refer the case without embarrassment. Fortunately, diagnostic aids exist. The electroencephalograph (p. 8), with expert interpretation, may point broadly, at least, to the site of disturbance; the concentration or sparsity of blood vessels in the pathological site may be visualized by arteriography; the distortion of the ventricular outline, or blockage of the ventricles, will show up in an X-ray plate if air is injected into them. These measures are confirmatory, the basis of diagnosis remains the intelligent and informed appraisal of the results of a careful examination of the patient's symptoms and performance, based on a sound knowledge of the areal divisions of the cortex and their functions.

Chapter Eight

STEREOTYPED MOTOR
MECHANISMS

THE ANALYSIS and synthesis of sensations by the cortex, and its organization and control of movement were grafted onto an already functioning mechanism of control, the difference being that the old motor system is automatic and rigid while the new control is voluntary and plastic. The old system took care of many mechanisms that are essential and are better if left automatic. In the interests of efficiency, this useful servant has been retained and occupies an important place in human neural organization. Unlike the cerebral cortex, it has no one location, but infiltrates the neuraxis and presents a number of sites and levels of concentration. We shall begin with the lowest level and add higher controls.

STATIC REACTIONS: THE RETICULAR SYSTEM. The lowest level in the brain stem is the reticular system of the bulb, pons and midbrain. Its higher levels are usually designated tegmental. Briefly, its occupies the space in the brain stem not preempted by named, organized tracts or nuclei. This places it in the central core. It is quite irregular and not perceptibly organized, so has been omitted from the diachrome reconstruction, but is labelled in the cross-sectional reconstructions, figures 48-51. The cells are scattered among coarse fibers of passage and vary greatly in size, but many are very large and stellate, and of the motor type. The nuclei may be somewhat arbitrarily divided into inferior reticular nucleus in the bulb, middle reticular in the pons and superior and deep tegmental in the midbrain. They send reticulospinal tracts downward, which are diffuse and may be crossed or uncrossed, but tend to concentrate in the ventral, outer part of the proprius fibers in the cord. The fibers end in relation to the motor cells of the spinal and cranial nerves.

The functions of the reticular nuclei are (1) to mediate vital reflexes depending on cranial nerves, mentioned on page 51; (2) to control and modify the spinal reflexes discussed on page 21; and (3) to mediate postural reflexes whose afferents pass over cranial nerves. Spinal reflexes enabling standing are rigid and unadaptive, the reticular influence permits a limb to be passively moved and then causes it to become rigid in the new position (lengthening and shortening reactions, respectively), and also enable a supporting limb to shift to a new spot when the body moves (shifting reaction). Examples of cranial nerve postural reflexes are the tonic neck reflexes which maintain the position of the head in relation to the body. When the head is put out of balance, proprioceptive fibers in the neck muscles, innervated by the accessory and higher cervicals, evoke a stronger contraction. Movements of the head will also produce compensatory alterations in the limbs, designed to maintain the center of gravity. All these will operate even if ocular and vestibular influences are excluded. The strength of the bulbar reticular reflexes is shown if they are released from higher control by sectioning the brain stem above the bulb. An extreme ridigity of the extensor muscles develops, decerebrate rigidity. The vestibules and the neck reflexes are factors in this caricature of normal standing. The object of this influence in the normal individual is to compensate for gravity by tonus, giving a neutral or zero status from which to make adjustments.

RIGHTING REACTIONS: THE RED NUCLEUS. At the upper end of the midbrain a large, rounded nuclear mass is located, in the middle of the tegmentum of either side, called the red nucleus (164: III; figs. 39, 51, 53) from the tint given by its vascularity in a cut of the fresh brain. Microscopically, the nucleus is seen to be composed of a large-celled portion which is small, and a small-celled portion which is large. The large-celled part is phylogenetically the older and is considered the more important in neurophysiology, as it sends out the rubrospinal tract (233: V M; figs. 51-48) which crosses below the nucleus and descends the brain stem, running laterally over the medial lemniscus in the pons, and under the spinal V tract in the bulb. It descends the entire length of the cord, placed ventral to the lateral corticospinal tract (figs. 5, 7), giving collaterals and terminals to motor neurons along the way, particularly to the axial musculature. The efferent connections of the small-celled part are poorly understood. They ascend to the cortex and descend to reticular nuclei and inferior olive. The great and obvious afferent connection of the red nucleus in general is the brachium conjunctivum (III: 165-167), which carries tonic impulses from the cerebellum, and loses most of its fibers here. A cortical influence is exercised by fibers from the forward part of area 6; striatal influence is mediated by pallido-rubral fibers (p. 125); ascending lemniscal connections enter; and the reception of vestibular impulses is physiologically obvious, probably through the MLF.

A quadruped whose brain stem has been sectioned above the red nucleus can right itself reflexly, but not if the section is lower. Data are brought in by the vestibular system, the proprioceptive system and the visual system. The mechanisms of the righting reflexes are to a large extent at the levels of the afferent and efferent nerves involved, but the control and selection is at the midbrain level. Asymmetric stimuli from the maculae of the labyrinth (p. 52) cause the head to be oriented correctly (labyrinth on head reflex). With the head rectified, but the body not, the stretched neck muscles set up a reflex contraction of the trunk musculature to rectify the body position (neck righting reflexes). When lying on the side, contact of the downward side can cause the head to be righted so the eyes are horizontal, even in absence of labyrinthine and visual information (body on head righting reflex). The body can be righted, in the presence of the red nuclei, if the feet are placed in contact with the ground. The visual righting reflexes are cortical, not rubral (red nucleus). In primates and man, lying on the side, the downward limbs are extended, the upward limbs are flexed and will grasp any object they contact (grasp reflex).

The reflex animal which we have been gradually constructing of spinal, reticular, rubral and visual reflexes is now able to stand with proper position of head, eyes looking straight forward, tonus distributed properly in the limbs; and it is able to regain this proper trim if it is disturbed. Gravity is compensated by extensor tonus; the heartbeat, respiration rate and other visceral activities are adjusted to the current needs, threats at the viscera countered by coughing, sneezing, blinking and the like, injuries to the body are corrected by withdrawal, with redistributed postural tonus. The individual, like a properly adjusted car with idling motor, is ready to respond efficiently and properly to any order the controlling will imposes on it. Being used to this ready state in ourselves and others, we can realize the importance of the accomplishment only when it is disturbed in patient or experimental animal.

QUALITY CONTROL: THE SUBTHALAMUS. There are three well-marked nuclei below the thalamus, between the hypothalamus in front and the midbrain tegmentum behind. They are the nigra (substantia nigra), subthalamic nucleus (of Luys) and the incerta (zona incerta). No one knows exactly what they have to do with control of stereotyped movement patterns, and they do not all do the same thing.

The nigra is a large, flattened cell mass covering the deep surface of the cerebral peduncle and placed rostral to the pontile nuclei (163: III; figs. 39, 51, 53). Its cells are large, dark and stellate, and contain a large amount of pigment. There are two distinct parts: a superficial diffuse-celled part, and a deeper, compact-celled part. Its connections are much under adjudication. There is a conspicuous and heavy connection with the lenticular nucleus, but it is not agreed which way it runs — probably it is lenticulo-nigral (162:

III; fig. 53). Many fibers from the cerebral peduncle veer off to enter it and some claim area 6 and others send to it, however many, possibly all the fibers return to the peduncle. A nigral descending tract is not well attested, but it apparently enters the tegmentum along with many descending pallidal fibers, to influence the reticular nuclei after being itself influenced by the lenticular nucleus, possibly adding a cortical factor. Its cells are degenerated in shaking palsy, along with those of the lenticular nucleus, so it is believed to have a smoothing effect on postural contraction discharges, like a fly-wheel on a gasoline engine, and other functions unknown.

The subthalamic nucleus (161: III; figs. 39, 53) is large and lens-shaped; it is placed just rostral to the nigra, and overlapping it. It lies in the high-way of the descending pallidal system, so it is difficult to be sure whether many fibers terminate there (160). It is believed that many of them do, and that the subthalamic nucleus relays them to the nigra. Another group is said to ascend to the pallidus, establishing a two-way connection. At any rate the subthalamus is not suspected of trafficking with the cerebral peduncle. In severe disease accompanied by purposeless, irregular, and uncontrollable movements (chorea or St. Vitus' dance, athetosis, hemiballism), the sub-thalamus is degenerated. Consequently, the nucleus is thought to act as a filter or damper to the free expression of momentary fluctuations in tonus potential, which the reflexes may develop.

The incerta is a thin, cellular lamina between the ventral thalamic nucleus above, and the subthalamic nucleus below (138: III M). Descending pallidal fibers form a hairpin turn around it medially, and pass through (as fascic-ulus thalamicus; 136: fig. 53); hence it is considered to be functionally cog-nate with the other nuclei of the subthalamic region.

Continuous laterally with the zona incerta, and forming a thin shell for the lateral aspect of the entire thalamus, is the reticular thalamic nucleus (not shown in illustrations). Virtually the entire intermediate thalamic radiation passes through it, and since corresponding fields of cells degenerate when the fibers traversing it die after experimental cortical lesions, it is thought to distribute widely to the cortex, but in a localized manner.

AUTOMATIC MOTOR PATTERNS: THE STRIATUM. The highest level, and most extensive representation, of the old motor system is in the cerebrum. It is the elaboration of the ventrolateral quadrant of the primitive cerebral hemisphere, the striatum (fig. 13). The other quadrants remain laminar; the striatum becomes massive. Attention was drawn to the relatively enormous development of the striatum in reptiles and birds (p. 29, figs. 14, 15), and the important role it plays in their motor and instinctual life. In man the striatum is still massive but is overshadowed by the neocortex and its connections.

The internal capsule might have skirted the striatum, instead it chose the shortest path and plunged through the middle of its prominent part. The path became a superhighway, and the caudate nucleus (126-128, dark ochre: II) was cut off from the remainder of the striatum, the lenticular nucleus (223-224: V), by the internal capsule (218-220: IV. See also fig. 55). That is, nearly so; fine, irregular cellular strands remain to attest the connection. They are shown (145), together with the spaces through which they pass in the capsule (IV). The internal capsule has a long, curved extent, and the caudate nucleus follows it all the way, except at the front (126, 223: fig. 55), where the two parts of the striatum are still in contact (matching II with V). The more massive head (126) gradually tapers to a body (127: II; fig. 52) and tail (128: II; figs. 54, 55), which follows the curve of the lateral ventricle through the temporal lobe (229: V), and terminates by joining the main mass basally where the sublenticular part of the capsule peters out.

The lenticular nucleus is shaped like a thick Brazil-nut, edge upward, truncated in front (223, 224: V; figs. 52, 53, 55). It is divided into medial and lateral segments, the pallidus (globus pallidus) (224, ochre) and putamen (223, dark ochre), respectively. This is a fundamental division, the putamen and caudate forming one indistinguishable unit functionally, and representing the neostriatum, while the pallidus alone forms the paleostriatum. The pallidus is apparently divided into inner and outer segments by a medullary lamina, but this is not functionally significant.

No fibers afferent to the striatal complex are discernible in myelin-stained sections of man except from the thalamus. Marchi studies after experimental lesions of the cortex show none. Yet when the suppressor areas of the cortex are stimulated, activity is detectible in the pallidus. So we must conclude that cortical influences either enter by totally unmyelinated collaterals or are relayed through the thalamus. The caudate-putamen unit contains many fascicles (225: V; figs. 52, 53), radially arranged with respect to the apex of the pallidus, and tapering to an end near the lateral surface. These arise from the large neurons of the caudate-putamen and end in contact with the pallidal neurons. In addition there are ten times as many tiny cells whose axons do not leave the caudate-putamen.*

The pallidus is dark with fibers and contains large cells whose axons pass down the brain stem. This is the counterpart of the lateral forebrain bundle of lower forms (fig. 13: l.f.b.). The descending course of the pallidal efferents is complex, but its comprehension can be simplified if one remembers that they end in a fair share of thalamus, hypothalamus, subthalamus, and tegmento-reticular, and that since all these are on the other side of the in-

---

* The recent introduction of the term "striatum" as a synonym for caudate-putamen forestalls its use for the functional unit: caudate-putamen-pallidus.

ternal capsule, they must first cross it. This they do in two main groups. The most frontal of them are able to form a loop around the edge of the capsule, forming the ansa lenticularis (135: II, V; fig. 52); the remainder must thread through in small strands, like hair in the teeth of a comb, the fascicularis lenticularis (227: V, 160, 162: III M; figs. 52, 53). From a different viewpoint there are also two groups: those running radially in the pallidus and out its apex are pallido-nigral (162: III; figs. 52, 53) (and/or nigro-pallidal plus nigro-putamenal); and those running concentrically within the pallidus. The concentric group includes two components, the ansa lenticularis, and the fasciculus thalamicus plus pallidobulbars (226 + 227: V; fig. 52). The latter component jogs sharply dorsally medial to the pallidus before crossing the capsule to form a part of the fascicularis lenticularis (the other part is the pallidonigral, which runs straight through the capsule).

The connections of the pallidus can be summarized as follows:
(See also chart on back of diachrome).

1. Ansa lenticularis fibers to and from the midline, lateralis dorsalis and ventralis anterior nuclei of the thalamus (135: II, V; figs. 39, 52).

2. The thalamic fasciculus, which runs medially under the incerta (so-called H2 field), turns around its medial edge (H), courses laterally over the incerta (H1), then into the ventral and lateral nuclei of the thalamus (227 in V, 136 in II; figs. 39, 53).

3. Hypothalamic fasciculus (227: fig. 52).

4. Supramammillary decussation to opposite side (with 162: fig. 52).

5. Pallido-subthalamics (160: III; fig. 53).

6. Pallido-rubrals (227: fig. 53).

7. Pallido-nigrals (162: III; fig. 53).

8. Pallido-tegmentals (226: V; fig. 51), pallido-bulbars and pallido-olivarys (140: figs. 51, 50, 49), apparently ending in the inferior olive, although they may be relayed to reach there. Pallido-spinals may or may not exist.

The connections with the thalamus are two way (136: II), not only with the ventralis anterior, but with the central nucleus (centre médian of Luys) (137: II; figs. 38, 39, 54), a distinctly visible rounded mass in the concavity of the arcuate nucleus.

Much effort has been spent to discover the function of the striatum. Comparison with lower forms indicates it has to do with control of movement, but in man this must be of an involuntary sort. Caudate-putamen removals or stimulations give no results. Pallidal removal produces muscular rigidity and inactivity. Pallidal stimulation may cause slow torsion movements. In experiments the danger is in involving the internal capsule.

In shaking palsy the pallidus degenerates. Shaking palsy patients, usually aged persons, show a coarse hand tremor, carry the body stiffly, have an expressionless "mask-like" face, and lack automatic associated movements, such as swinging the arms while walking. These may be considered the negative of the function of the pallidus.

In chorea and athetosis the caudate-putamen may be degenerated, or else structures which are thought to connect with it via the thalamus (brachium conjunctivum, frontal cortex), although here the case is not so clear. The interpretation is that the caudate-putamen and its afferents control the pallidus, which then hyperfunctions, giving a hypermotility of a purposeless sort.

If the pallidal efferents are severed at operation the tremor and rigidity are decreased or lost in a large proportion of victims of shaking palsy. Some neurosurgeons utilize a stereotaxic machine (p. 8) built to accommodate the human brain, either using x-ray as a guide, or using analogous measurements from sectioned heads. A few rely on dead reckoning from surface measurements. Undercutting of the premotor cortex is said to reduce athetotic movements with little functional loss.

There are two terms which have been used with such inexact meanings that they have caused confusion. One is basal ganglia, which may mean the striatal complex, with or without amygdala, subthalamus and even thalamus. The other is "extrapyramidal system," which is convenient if applied to the structures in this chapter as a group, but it often includes vestibulospinal and tectospinal tracts, and is even extended to include area 6, 8, and any other motor control systems except the pyramidal, i.e., is synonymous with "non-pyramidal motor."

The effects on the stretch reflex of different parts of the nonpyramidal motor system have been studied. The inferior reticular nucleus suppresses the reflex when stimulated, so is called the bulbar suppressor center. The suppressor strip regions of the cortex, especially in area 6, also inhibit the stretch reflex, and the path can be shown to pass down the capsule and pyramid into the bulb.

Higher in the brain stem is a center, stimulation of which facilitates the stretch reflex. This in turn is acted upon by the pallidus and subthalamus. In the balance of these two factors lies the proper distribution of tonus requisite for normal activity.

The classic ascending paths emit collaterals in the central core of the bulb, these converge on nonspecific ascending paths in the tegmentum which apparently reach the medial, submedial and reticular nuclei of the thalamus, whereby they induce an alerting or awakening reaction. In turn this excitement is relayed into widespread regions of the cerebral cortex, particularly frontally. The entire system forms the nonspecific reticular activating system.

Chapter Nine

CEREBELLAR REGULATION
OF TONUS

THE ROLE of the cerebellum in movement is like that of a ground crew which makes a passenger service possible. The cerebellum never causes a movement, but willed movements would not get far, and posture and stance would be poorly adapted if the cerebellum did not get in its work. The cerebellum receives information from every source, sensory and motor, of possible usefulness to it, and transforms this into adjustments of tonus in muscles maintaining posture or performing movements.

The afferent streams to the cerebellum are three, and they are very unequal in man. The vestibular nerve gives a small connection to the caudal extreme of the cerebellum; the spinocerebellar tracts, supported by the other components of the restiform body, discharge in the anterior and medial regions; while the cerebral cortex connects through the brachium pontis and the striatum through the inferior olives, to the massive lateral lobes.

The cerebellar cortex does a magnificent job of combining the signals of each of the three categories and turning out from the first group adjustments of equilibrium (vestibulocerebellum), from the second group adjustments of posture (somatocerebellum), and from the third group adjustments of movement (cerebro-cerebellum).

The cerebellum developed in fishes because the vestibular sense needed a place where it could interchange with the lateral line system of nerves (fig. 10), which fish use to perceive depth, waves and near objects, but which are completely lost (not transformed) on assumption of terrestrial life. So in fishes the vestibulocerebellum is nearly or quite the entirety of the cerebellum, and it may be very large. The cerebellum developed at the junction of bulb and midbrain (figs. 8, 11) because vestibular and lateral line nerves entered at this level, and is dorsal because it was served by the special somatic sensory component. Fusion in the midline was essential to its overall influence. In reptiles and birds, posture is complicated by the influence of gravity and development of the limbs, and movements are more complex, so somatic and striatal connections are developed, which enter in front of the earlier portion, establishing the somatocerebellum. Finally, the development of cortical control of movement in the mammals demanded cooperation of the cerebellum, so a connection was made which swings around the brain stem as the brachium pontis, and causes great expansion of its lateral lobes. This is the cerebrocerebellum.

Thus, there are three stages in the development of the cerebellum. The stages are correlated with new functional demands, and because the newly-developed tracts reach separate parts of the cerebellum, specific regions are associated with specific functions, connections and stages of development. These stages are usually called archi- (ancient), paleo- (antique), and neo- (new) cerebellum, but we prefer the functional designations vestibulo-, somato- and cerebrocerebellum.

CEREBELLAR CORTEX. To achieve its ends, the cerebellum must receive information from every possibly useful source, and it must make as complete a mixture of these as possible to effect an algebraic addition of them. This it attains by a clever arrangement of neurons, the cerebellar cortex. Now, the cerebellar cortex should not be thought of as at all analogous to the cerebral cortex; in fact, the reverse, for the cerebral cortex is organized into units, small and large, to make specificity possible. The scheme was hit upon in the earliest form, and has scarcely been altered in man.

The cerebellar cortex is disposed in long parallel folia (fig. 44), which cover the laminae of the fibers which serve it, and which stand out from a central medullary core of fiber tracts. Samples taken anywhere are identical in structure. (Consult figure 43 in connection with the following description.) Anatomically there are apparently two layers, although functionally they are one. The outer layer, of uniform thickness, is poor in cells but rich in synapses. In the inner layer, thicker in the folial crests than in their troughs, are packed some of the most minute cells in the brain, the granule cells. Looking closer, one sees at the junction of the two layers a single line of very large cells, spaced like a row of onions in a garden, the Purkinje cells (a). All the other elements feed into the Purkinje cells, whose axons

→

Figure 43. TRANSPARENT STEREOGRAM OF A BIT OF CEREBELLAR CORTEX SHOWING THE VARIOUS NEURONAL ELEMENTS AND THEIR CONNECTIONS

For explanation see text.

a. Purkinje cell	g. Granular cell axon
b. Purkinje axon	h. Tangential branch of g
c. Purkinje collateral	j. Basket cell
d. Climbing fiber	k. Basket cell axon
e. Mossy fiber	m. Stellate cell
f. Granular cell	n. Golgi type II cell

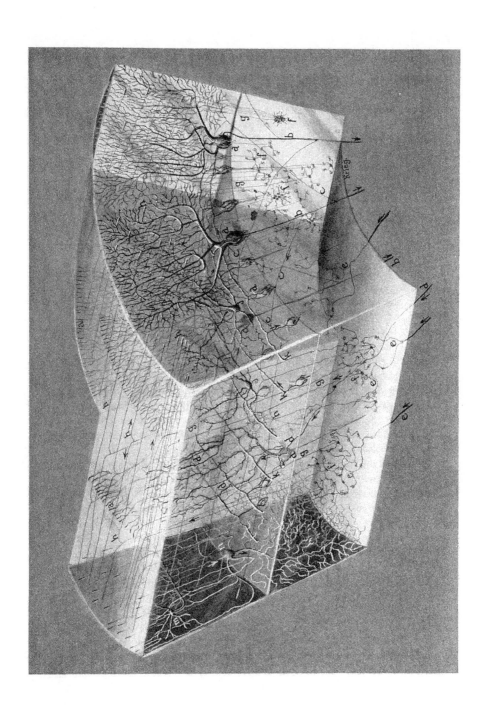

alone leave the cortex. This cell is the most remarkable of appearance in the body. One or several thick dendrites branch within the outer layer, but strictly in one plane, that running across the folium, and the branches are so numerous as nearly to fill the interspaces, so the neuron closely resembles a branch of arbor-vitae cedar. The axon (b) comes from the bottom of the cell and traverses the granule-cell layer, but sends back collaterals (c), which branch just under the Purkinje cell bodies, forming the most elaborate plexus of collaterals observable anywhere. Thus, the Purkinje cells gather impulses from a wide space, and they in turn excite numerous other Purkinje cells.

The Purkinje cells receive a variety of axons. The simplest are the climbing fibers (d) of some of the afferent tracts, no one knows just which. A single fiber reaches a Purkinje cell body and branches with its dendrites, like a vine on a tree.

A longer circuit is through the granule cells. Other afferents (e) wander like moss among the granular cells (f), for long distances, making contacts along the way, and terminally, with the granular cell dendrites, which are tiny hands on the several radiating arms from the cell body. The cells and endings are in flecks or clumps. Their axons (g) enter the outer layer and bifurcate like a T into two tenuous filaments (h), which run long distances parallel to the folia and each other, threading between the Purkinje-cell dendrites like wires between telegraph poles.

The bodies of basket cells (j) lie in the basal part of the outer layer, and their dendrites spread out diffusely in the same layer. Their axons (k) run parallel to the surface, but in a direction across the folia. Opposite each Purkinje cell body they send out a branch which surrounds it with twigs like the wickerwork around a demijohn. Other cell types are the stellate cell (m) in the molecular layer, similar to the basket cell but less differentiated; and the Golgi type II cell (n), with a dendritic spray in the outer layer and a similar axonal spray in the granular layer. It is evident that the neuronal pattern is adapted for the widest possible dissemination of any impulse. The three planes of space are criss-crossed by a latticework of neuron terminations: the granule cell axons parallel to the laminar axis, the basket cell axons running across the folia, the Purkinje dendrites perpendicular to the surface. Any localization observable in the cerebellar cortex is only the result of physical limitations on a distributive system of maximum efficiency. Although anatomically several layered, functionally there is a single layer, since all neurons channel into the Purkinje cells. Purkinje cell axons pass either to the central cerebellar nuclei on their way out, or to other folia.

TOPOGRAPHY. In a midline section the cortex of the cerebellum seems to be curled up like a grub-worm (82-84: I). That is why the medial zone is called the vermis. Behind and below, the vermis is ridge-like and separated by clefts from the prominent lateral lobes which form a pair of buttocks enclosing it (fig. 44). Above, the vermis is directly continuous with the lateral lobes (III, V; fig. 44). The vermis is deeply incised into lobules (117-125: I L), and the lobules notched into folia. Two perseverating sulci divide the cerebellum into three lobes: anterior (82, pink: I M), middle (83, yellow) and posterior (84, pale blue). These three lobes roughly correspond, respectively, with somatocerebellum, cerebrocerebellum, and vestibulocerebellum.

The lobules are continuous to a variable degree with corresponding divisions of the lateral lobe. Each vermis lobule has been given a fanciful name by early anatomists, and the best thing to do is learn the sequence by rote and have done with it. From before backwards, they are (see I L): lingula (117), central lobule (118), culmen (119), declive (120), folium (121), tuber (122), pyramis (123), uvula (124), nodule (125). A recent comparative study demonstrates their unity among vertebrates, traces them down the phylum and proposes Roman numeral designations I-X as follows: I—lingula; II, III—centralis; IV, V—culmen; VI—declive; VII—tuber; VIII—pyramis; IX—uvula; X—nodule (fig. 44). Following the lateral relations (fig. 44): Lingula stops close to the median line, central lobule and culmen form anterior quadrangular lobule. All these form the anterior lobe. Declive expands greatly to form posterior quadrangular lobule (simple lobule). Folium expands into superior semilunar lobule. Tuber begins the inferior aspect of the cerebellum and forms inferior semilunar and slender lobules. Pyramis forms biventral lobule and tonsil (173; III M). The multiplication of folia laterally causes an increasing fullness so they bulge downward and finally medially, like the pages of a book which has been wet. This group is the middle lobe and comprises the greater bulk of the cerebellum. Uvula has no lateral representative. Nodule is connected to its flocculus (256: V L) by a thin lamina (posterior medullary velum). These form the posterior lobe. The flocculonodular lobe is the earliest to separate from the body of the cerebellum, and constitutes a distinct unit, which receives the vestibular connections.

COLLECTION OF DATA BY THE CEREBELLUM. Vestibular data. The vestibular nerve sends primary fibers of its ascending root, and secondary fibers which have synapsed in the superior vestibular nucleus (203: IV; fig. 49), to the fastigial nucleus of the cerebellum (p. 54). A long vestibular bundle curves over the restiform body, to the flocculus (204: IV; fig. 49), and carries fibers back to the vestibular nuclei; together they form the peduncle of the flocculus.

Muscle-joint data: The spinocerebellar tracts of the spinal cord were described on pages 19 and 20, and illustrated in figures 6 and 7. In the bulb they are located near the periphery dorsolaterally (240, 241: V M; figs. 45, 48). In the upper bulb the dorsal tract (240: V M; fig. 49) climbs gradually more dorsally and joins the other components of the restiform body. The ventral tract joins the restiform body from below, but remains medial (241: V M; figs. 45, 49, 50), and enters the cerebellum by running dorsally outside of the brachium conjunctivum (167: III; fig. 50). They are distributed to the vermis (170: III), more specifically to centralis, culmen, pyramis, uvula. The ventral tract is more restricted in termination than the dorsal, concentrating in the medial part of the central lobule and culmen. The dorsal spinocerebellar tract relays proprioceptive impulses from trunk and lower limbs of the same side, the ventral tract conducts from all levels, is mostly crossed, but recrosses in the cerebellum. When the spinocerebellar tracts are cut, loss of muscular tone and muscular incoordination result, but there is no loss of conscious proprioception or movement. These symptoms are similar to those of destruction of the somatocerebellum, and the distribution of the spinocerebellar tracts may be taken as coinciding with the extent of the somatocerebellum. It receives other fibers, however.

The external cuneate fibers. The fibers from the upper limbs that correspond to those joining the dorsal spinocerebellar system run in the dorsal funiculus as primary fibers (238: V M) as far as the middle of the bulb. Here is located a nucleus, lateral to the cuneate nucleus, having cells like those of Clarke's column, the external cuneate nucleus (239: V M; fig. 48). From here, axons ascend with the restiform body of the same side and end in the anterior lobe.

STRIATAL DATA: THE OLIVOCEREBELLAR FIBERS. The inferior olivary nuclei are the most conspicuous feature in sections of the upper half of the bulb. The main nucleus of the group occupies a large share of its central part. It is ovoid in general outline (255: V M; fig. 28) and shaped like a crumpled purse. In sections it is seen to be widely open medially and abundantly folded (figs. 48, 49). Its cells form separate baskets or knots of dendrites. Their axons, the olivocerebellar fibers, emerge from its hilus, cross the medial lemnisci of both sides, and swing laterally and dorsally on the opposite side

→

Figure 44. ASSEMBLY DIAGRAM OF THE LOBULES OF THE CEREBELLUM

In the center is the complete cerebellum in anterior view. The yoke-shaped cut surface represents its attachment to the brain stem; with brachium pontis (middle peduncle) below and laterally, restiform body (inferior peduncle) below and medially, and brachium conjunctivum (superior peduncle) above. The cerebellar cortex is made up of a system of regularly arranged folia grouped into lobules. Each lobule is separately displayed and named on the right-hand side.

The name on the medial aspect of most of the lobules indicates the lobule of the vermis to which that lobule contributes

On the left-hand side the lobules are grouped according to several systems. In italics are the functional designations: somato-, cerebro-, and vestibulocerebellum. In lower case are the designations used classically in comparative anatomy. On the right-hand side the Roman numerals are the newer comparative designations of Larsell.

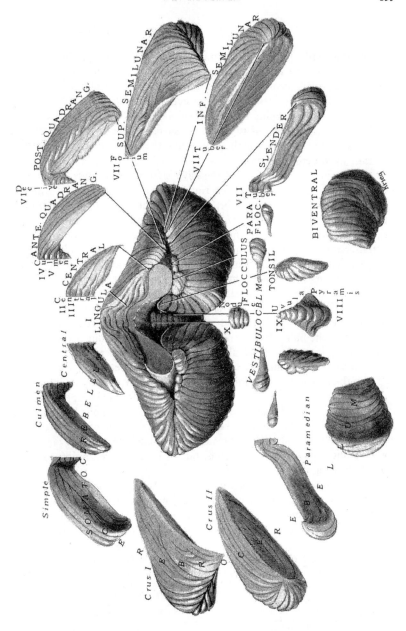

to join the restiform body on its medial aspect (255: V M; fig. 45). Medial to the main olivary nucleus is a flat plate of cells, resembling it in type and origin, the medial accessory olive (255m: fig. 48); while above it is the dorsal accessory olive. The dorsal accessory olive projects to the somatocerebellum. The main nucleus supplies the cerebrocerebellum, its dorsal leaf to the upper surface and its ventral leaf to the lower surface. The medial accessory olive projects to the vestibulocerebellum. Thus the inferior olive is distributed to the entire cerebellar cortex.

The main afferent supply of the olive is from the pallidus, as the pallido-olivary tract (226: V M), which runs down the middle of the tegmentum (figs. 51, 50, 49) and forms a sort of fleece around the olive. Rubro-olivary, tegmento-olivary, spino-olivary, and other extrapyramidal connections also exist. Through this system the cerebellum is informed of the doings of the extrapyramidal motor system. Localized olivary lesions indicate this, as exaggerated tonus, rigidity, incoordination, and inaccurate movements result—mirroring the striatal contribution to movement.

Arcuate nuclei are irregular cell groups daubed on the pyramidal tracts (fig. 48). They send axons to the restiform body of the same side by skirting the outside of the bulb, and to the opposite side by traversing the median line of the bulb, then laterally on the floor of the fourth ventricle as the striae medullares. There are also contributions to the restiform body from the lateral reticular nuclei and other tegmental structures.

All the above tracts combine to form the restiform body or inferior cerebellar peduncle, which forms a prominent shoulder in the upper bulb (figs. 45, 49) then turns dorsally into the cerebellum (242, light brown: V M). In the medullary core of the cerebellum it lies in front of and lateral to the dentate nucleus (168: III) and medial to the brachium pontis (246: V); and is distributed chiefly to the vermis, but in lesser concentration to all parts of the cerebellum.

CEREBRAL DATA: THE BRACHIUM PONTIS. In the human brain, by far the largest cerebellar connection is the brachium pontis, or middle cerebellar peduncle, consisting exclusively of axons of the cells of the pontile nuclei of the opposite side. The nuclei (243: V M; fig. 50) occupy the irregular spaces

→

Figure 45. PHANTOM OF THE BRAIN STEM IN DORSAL VIEW SHOWING THE COURSE OF FIBERS COMPOSING THE CEREBELLAR BRACHIA

C. Attachment of chorioid roof of fourth ventricle
D. Dorsal spinocerebellar tract
E. External cuneate nucleus
F. Frontopontile fascicles
I. Restiform body (inferior peduncle)
L. Lateral corticospinal tract
M. Brachium pontis (middle peduncle)
O. Inferior olive
O-C. Olivocerebellar fibers

Pa. Parietopontile fibers
P-C. Pontocerebellar fibers
Pe. Cerebral peduncle
Po. Pontile nuclei
Py. Pyramidal tract
S. Brachium conjunctivum (superior peduncle)
T. Temporo- and occipitopontile fibers
V. Ventral corticospinal tract

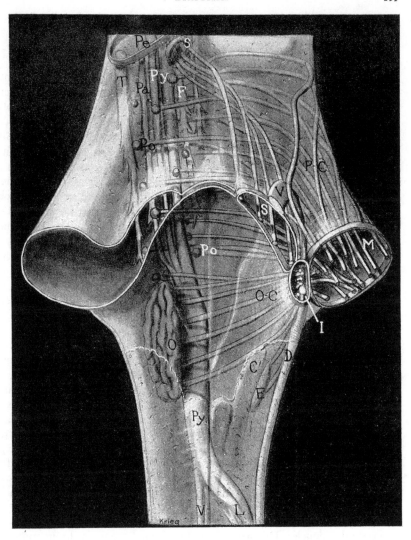

in the pontile bulge, left by the descending cerebral fibers and the accumulation of the pontocerebellar fibers, which are from their cells. The axons cross the midline in small fascicles (fig. 45) except superficially and deeply, where they form more or less compact laminae (244, 245: V M; fig. 50). At the lateral face of the nuclear zone they form a massive bundle (fig. 50) directed laterally and dorsally toward the lateral lobes of the cerebellum (246: V; fig. 45). The arrangement of fibers is an orderly one, as shown in figure 45, and the fibers to the various folia curve off in a direct manner, the deeper, rostral fibers to the anterior lobe, the lateral fibers of superficial origin to lateral stations on the cerebellum, and so forth (247: V M). The upper surfaces of the lateral lobes (posterior quadrangular and superior semilunar lobules) except medially, receive the best supply, the lateral parts of the anterior lobe somewhat less, the inferior semilunar lobe only a few, and the biventral lobule, tonsil and vermis few or none. The omitted regions show numerous arcuate fibers forming a cerebellar associational cortex, although this is not generally known. Except for the olivocerebellar fibers, virtually no other connections reach the lateral lobes of the cerebellum, so their functional nature must depend on what the pontile nuceli receive.

As described in chapter VII, many areas of the cerebral cortex send fibers through the internal capsule and peduncle to terminate in the pontile nuclei. In fact, virtually only the area 4 fibers run the gamut. The chief contributor is area 6, which conveys information on the organization of motor acts (and, as we shall see, is in turn influenced by the cerebellum). Other strong contributions come from the parietal region, particularly associative areas 7 and 5, less from 2 and 1, and not at all from 3. These may be responsible for the cerebellar activity noted after tactile stimulation. There are definite occipitopontile connections and these are the sole source of cerebellar activity after optic stimuli. Some parts of the temporal cortex connect with pontile nuclei, but auditory stimuli are routed to the cerebellum only through the inferior colliculus.

All of the corticopontile fibers are included with the internal capsule and cerebral peduncle. The premotor and prefrontal group, in the anterior limb of the internal capsule (218: IV), from the medial fifth of the cerebral peduncle (187, 188: IV; figs. 50, 51) and end among the medial pontile cells. The parieto-, temporo-, and occipito-pontile fibers are arranged in that order in the lateral two-fifths of the internal capsule (189: IV; figs. 50, 51), and end among the lateral cells of the pons.

In summary, the cerebellum is informed of the state of the sensory and motor nervous systems at every instant.

DISTRIBUTION OF TONUS TO MUSCLES. These data are put through the cortical processing as described, and efferent messages pass out the Purkinje axons. It is a rule that no cortical axons leave the cerebellum, although a few of the medial ones probably do. Instead, they pass to one of four central or tectal nuclei in the middle of the medullary center, and these, in turn, send out axons to the extracerebellar structures. These lie side by side at the center of the cerebellum, and are very unequal in size. From medial to lateral they are named fastigial, globose, emboliform and dentate nuclei.

The fastigial nucleus (142: II) receives vestibular connections and cortical axons from the somatocerebellum. It sends out two sets of fibers: the fastigiobulbars, which run back along the restiform body to end in the reticular nuclei of the brain stem; and the hook bundle, which crosses above the tectal nuclei, curves lateral to the brachium conjunctivum, then medial to the restiform body to end among the vestibular and reticular nuclei (143: II; fig. 50). These tracts are important in maintaining equilibrium.

The emboliform nucleus (169: III) receives its afferents from the somatocerebellum (171: III). Its efferents form the most medial fibers of the brachium conjunctivum (167-165: III; fig. 45) and end in the large-celled division of the red nucleus (164). Thus the emboliform nucleus is able to influence postural tonus through the rubrospinal tract. If fastigial and emboliform nuclei are destroyed, or the anterior lobe is removed, the stretch reflexes, lacking cerebellar inhibition, produce great rigidity and spasticity. If they are stimulated, postural tone is lost and the standing animal collapses. The connections of the small globose nucleus are similar to those of the emboliform.

The dentate nucleus (168: III) is the most lateral of the four. It is much the largest and resembles the inferior olive in being hollow, having a corrugated outline and opening medially. It is, however, more pouch-like. Into it run the great proportion of the Purkinje axons (172: III), all except those cited as entering the other three. It is thus properly considered a part of the cerebrocerebellum or the neocerebellum. Its cells send their axons toward the central cavity, which they fill. As it leaves the dentate nucleus, the brachium conjunctivum, or superior cerebellar peduncle, runs forward medial to the brachium pontis and, at first, above the brain stem, forms the walls of the upper part of the fourth ventricle (167: III; figs. 45, 50), crosses massively in the midbrain tegmentum (166) and continues to run forward (165) a short way until it reaches the red nucleus (164), where it breaks into fascicles and loses many fibers to the parvocellular portion. About a third of the fibers continue in the thalamus to end in the nucleus ventralis ventralis (130: II L, III; figs. 37, 53). Axons arising here enter the forward part of the intermediate thalamic radiation (130, brown: III), join the area 6 and 4 parts of the corona radiata and end abundantly in area 6, less so in area 4.

Thus a cerebellar influence is stamped on movements as they are being organized (and on other cortical regions, to the uncertain degree they are supplied by the parvocellular red nucleus). Injury to any part of the dentate nucleus or branchium conjunctivum will give the symptoms of cerebrocerebellar injury, which will be discussed in the next section.

We have seen there is ample anatomical counterpart for the three functional divisions of the cerebellum, but that there is a certain amount of overlap at the edges. The tendency of the cerebellum is to blend all connections which enter it, but over such a large area this is impossible, and there are local differences in functional concentration, due to the natural tendency of tracts to end where they strike the cortex.

When anesthetics are used to suppress cortical spread of impulse, a rough topological representation of the body can be adduced. The body is represented on the superior vermis in reverse order, the tail in front at the lingula, the head behind at the culmen; the limbs are represented laterally, the hind limb in front, the front limb behind, but everything is ipsilateral. There is a separate localization map in the lateral lobe representing the pons. Under ordinary experimental and clinical conditions, localization, except for laterality, is not detectible. Stimulation of the somatocerebellar cortex in animals sometimes leads to movement, but considering the elaborate background of tonus, this is not surprising. Cerebrocerebellar stimulation is without phasic effect.

FUNCTIONS. When the vestibulocerebellum of one side is removed in experimental animals, the subject staggers to the opposite side. Children with tumors of the nodule sway, stagger and fall. If the somatocerebellum is removed in animals (human involvements being rare), the postural muscles become rigid. Correspondingly, when it is stimulated, postural tonus is inhibited, and the animal tumbles in a heap. Functions of the cerebrocerebellum are best seen in man because in no animal, even higher apes, is it as well developed, and the subjective and fine-grained differences are better observed in a subject with whom the examiner can communicate. Because of the size and prominence of the lateral lobes in man, they are most usually and the most heavily affected; in fact, their affliction is tantamount to the clinical cerebellar syndrome. The symptoms are exhibited on the side of the lesion, and there may be a tendency to deviate to the same side. All symptoms are seen only when movement is attempted, and especially when the patient is standing. There is a general quality of weakness (asthenia), especially of the proximal muscles. There is a loss of tonus (atonia), so the limbs assume unusual positions, and tend to be limp. Most characteristically there is loss of muscular coordination. This may be illustrated in various ways. The movement may fail of its intended mark (dysmetria), as when the patient attempts to touch the nose with the finger-tip when the eyes are shut. The ability to adjust the tonic state quickly is lost as, when the patient attempts to flex his arm when held by the examiner, he hits his face when the arm is released (rebound). The several components of a movement are broken (decomposition of movement), giving a result like performing the manual of arms "by the numbers." The ability to perform rapidly alternating movements is lost (adiodochokinesis. a = without, diodocho = duplicity, kinesis = movement) as when the hands are rapidly changed from palms up to palms down. There is a coarse tremor when reaching (intention tremor). Speech may be slurred, monotonus, rhythmical. All of these symptoms combine into cerebellar ataxia, which differs from ataxia after lesions of the posterior funiculi of the cord in that there is no sensory loss.

To sum up, the vestibulocerebellum distributes tonus to the muscles to maintain balance; the somatocerebellum regulates tone for the postural mechanism; while the cerebrocerebellum regulates muscular tone in motion, facilitating smooth movements. All in all, the cerebellum coordinates and regulates the muscles, but is not itself a prime mover.

Chapter Ten

VISCERAL REGULATION

LAN OF THE AUTONOMIC NERVOUS SYSTEM. First of all, let it be clear that the viscera in general perform their intrinsic functions without nervous control. Experimental animals have lived long periods in the laboratory and even reproduced, after their autonomic nervous connections have been severed. A constant neural tippling by the viscera is not necessary. Such animals, however, could not survive in nature, because their organs cannot adjust their response to rigors of the environment, pursuit, feasting and famine. The sympathetic system adjusts the organism to all forms of excitement or stress — fear, rage, combat, flight, pursuit. The parasympathetic system functions in quiet, vegetative states. Nearly all organs have a double innervation, and one of these exerts a stimulating effect, the other an inhibitory effect. The sign or nature of the effect can be divined for any organ by asking oneself whether that organ's activity should be exalted or diminished during excitement.

The organization and pathways of autonomic nerve supply are our chief concern here, however. The connections of the sympathetic system with the spinal cord, the plan of distribution of its neurons, and the organization of its terminal plexuses will be detailed. Most of the parasympathetic system has been described under the heading of visceral efferent system (p. 47), but its sacral part remains to be described. It should be borne in mind that the cranial division of the parasympathetic is the functional counterpart of the greater portion of the sympathetic, and shares its terminal plexuses in complete intimacy.

PREGANGLIONIC SYMPATHETIC NEURON. In the spinal cord, where sensory and motor cell columns meet, is a little ridge composed of sympathetic neuron cell bodies, the lateral cell column (figs. 4, 46). It extends only throughout the thoracic and upper two lumbar segments, and its efferent axons pass out with the spinal nerves, but only in segments T1 to L2. As the spinal nerves leave the vertebral canal, these neurons depart from the main trunk and run ventrally alongside the vertebra as the white ramus communicans (white because myelinated). Each enters a separate small paravertebral sympathetic ganglion. The series of ganglia are connected like a chain by nerve bundles, forming on each side the sympathetic trunk (figs. 31, 46, 47). Some preganglionic fibers synapse with the nerve cells of the postganglionic neurons in their corresponding ganglia. Others of them ascend the chain, which is continued upward through the cervical region, and still others descend the chain, which is continued downward through the sacral region. In fact, the cervical and sacral sympathetic ganglia can be supplied only by ascending or descending fibers, respectively, as there are no white rami from their own segments. Thus there is considerable up and down traffic of the preganglionic fibers.

POSTGANGLIONIC NEURON. The paravertebral sympathetic ganglia are like the spinal ganglia in that the cell bodies are surrounded by capsules; however, here the cells are multipolar. Dendrites are of two classes: short dendrites which form a glomerulus within the capsule, and dendrites which run long distances in the ganglion, often in bundles. Both types synapse with the preganglionic axons.

Axons from sympathetic ganglia are called postganglionic fibers. They are unmyelinated; the sympathetic ganglion has not the power to give a myelin sheath to its fibers. Those to the viscera of thorax, abdomen and pelvis continue along the walls of the arteries which supply the several viscera, in the form of plexuses.

The visceral structures in the body wall and limbs, such as arteries and sweat glands, must receive a postganglionic supply, too. Each ganglion sends a gray ramus communicans (gray because unmyelinated) to the spinal nerve with which it is associated, and the sympathetic fibers are distributed by it, although they tend to run rather long distances with the arteries. Thus, there is a grey ramus communicans to each and every spinal nerve, but a white ramus only from nerves T1-L2. The sympathetic supply of the head deserves special mention. The preganglionic fibers come out of the cord by the first thoracic white ramus, pass up the cervical portion of the sympathetic trunk. Instead of the expected eight ganglia, there are only three, due to fusion. The superior cervical ganglion is by far the largest, being formed of ganglia for nerves C1-C4. From its upper end, plexuses pass along the internal and external carotid arteries, and send twigs through each of the four parasympathetic ganglia of the head, without synapsing, and sympathetic and parasympathetics are distributed together to the organs they supply (fig. 30, 47).

THE PARASYMPATHETIC SYSTEM is divided into two portions, cranial and sacral. As described on pages 47 to 50, the cranial preganglionic neurons pass out of the brain stem with nerves III, VII, IX and X, and synapse in ciliary, sphenopalatine, otic and submaxillary ganglia. These resemble sympathetic ganglia in their structure. The postganglionic fibers generally hitch a ride on passing nerves to be distributed to the glands and smooth muscles of the head.

The vagus nerves furnish parasympathetic innervation to the viscera of thorax and abdomen, but not to those of the pelvis. The preganglionic fiber runs the entire length of the nerve, and the postganglionic cell bodies are in numerous microscopic ganglia, containing only a few cells each, within the walls of the organ itself, linked together as a plexus (fig. 46).

In the alimentary tract the ganglia are in two positions, between the

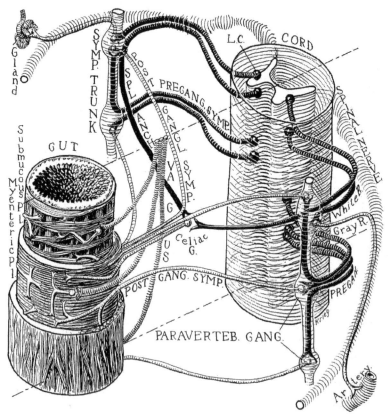

Figure 46. STEREOGRAM OF CONNECTIONS OF AUTONOMIC NERVOUS SYSTEM

For explanation see text.

Preganglionic sympathetic neurons are black, vagal (preganglionic parasympathetic) neurons are hatched, postganglionic neurons of both types are white. The spinal nerve is represented in phantom.

inner and outer muscular layers, the myenteric plexus; and in the submucosa, the submucous plexus. Thus the postganglionic fibers are very short. The postganglionic sympathetic fibers contribute to the plexus, and accompany parasympathetic fibers to their terminations, but do not, of course, synapse in the ganglia of the plexuses, since they are already postganglionic.

The vagus nerve does not supply the viscera of the pelvis. Instead, sacral segments 3 and 4 contain preganglionic cell bodies whose axons pass out with sacral nerves 3 and 4, leave by white rami and join the pelvic parasympathetic ganglia at the sides of rectum and prostate or vagina. They are joined by the sympathetic visceral branches, some arising directly from the sacral sympathetic ganglia, and others arising in the lumbar ganglia, and joining into the hypogastric nerve, which runs over the back of the pelvic rim and splits to form the pelvic plexus.

There are no parasympathetic grey rami to the spinal nerves, and the visceral structures in predominantly somatic regions have no parasympathetic nerve supply.

INNERVATION OF THE SPECIFIC ORGANS. A brief summary of the autonomic nerve supply of the main organs will now be given. Compare the anatomical descriptions with figure 47.

HEART. The sympathetic supply is from superior, middle and inferior sympathetic ganglia, each giving off a cardiac sympathetic nerve (fig. 30). The vagal supply is through superior and inferior cardiac nerves in the neck. All join to form the cardiac plexus around the arch of the aorta. The heart is accelerated by the sympathetic, slowed by the vagus.

RESPIRATORY SYSTEM. The sympathetic supply of the lungs is by visceral branches from paravertebral ganglia T3-T5. Vagal supply is from a plexus of branches as the trunk runs past the lung hilus. These combine to form the pulmonary plexus, which follows the branching of the bronchioles. The bronchi are contracted by the vagus. In bronchial asthma this occurs to a distressing degree. Drugs, such as adrenalin and ephedrin, which stimulate the sympathetic, are in general use to restore the normal bronchial caliber.

DIGESTIVE SYSTEM. Since the entire alimentary tract from esophagus through rectum is supplied with autonomic fibers, it may be said that all sympathetic ganglia contribute to its supply. Thoracic ganglia 5-12 send off large visceral branches which combine to form three splanchnic nerves (fig. 47), composed of preganglionic fibers which have passed through the paravertebral ganglia without synapsing (fig. 46). They synapse in the visceral ganglia, the chief of which is the celiac. From here, postganglionic fibers pass out along the branches of the celiac artery to stomach, liver, pancreas and small intestine. At the root of the superior mesenteric artery is the superior mesenteric ganglion, whose fibers help to supply the small intestine.

Similarly located is the inferior mesenteric ganglion for the non-pelvic colon. Lumbar visceral branches contribute to the supply of the colon.

Parasympathetic supply of the digestive tube is from the vagus, which runs down the sides of the esophagus and forms a plexus around the stomach (fig. 47). Twigs continue to liver, pancreas, small intestine and colon, all preganglionic. The postganglionic cells are in myenteric and submucous plexuses. The sigmoid colon and rectum are supplied by the pelvic plexus, receiving sympathetic roots from lumbar and parasympathetic roots from S3 and 4.

The vagus stimulates peristalsis and digestion, the sympathetic inhibits it. This is why states of excitement or anxiety are not conducive to digestion. The vagus also enhances the secretion of digestive enzymes, and, perhaps, of insulin, and favors storage of sugar in the liver. The sympathetic arrests peristalsis, mobilizes sugar stored in the liver and creates a blood reservoir in the splanchnic vessels. In duodenal ulcer the vagus is sometimes sectioned to rest the stomach and decrease enzyme secretion.

URINARY SYSTEM. The sympathetic supply of the kidney is from T10-12 through the celiac plexus. Its parasympathetic supply is derived from the vagus. The terminals are to the vessels, apparently not to the tubules, but in stress the sympathetic has the power of contracting the glomerular efferent arteriole, decreasing renal blood flow and, consequently, urinary secretion.

The bladder is supplied with sympathetic fibers from the upper lumbar roots and from S3, 4 on the parasympathetic side. Afferent impulses from the full bladder through the sacral visceral nerves initiate a reflex relaxation of the internal sphincter and contraction of the bladder muscle, but emptying will not occur unless the striated external sphincter is voluntarily relaxed. When the cord is severed, the emptying reflex of the bladder becomes automatic, but the bladder does not empty completely.

GENERATIVE SYSTEM. Male. The testis develops in the middle of the abdomen and carries its autonomic supply with it, from T10 to L1. The prostate and penis share the pelvic plexus. Seminal secretion is not much influenced by nervous control, but erection and ejaculation are. The erection reflex is effected by relaxation of the arteries of the penis, permitting a filling of the cavernous spaces which in turn constricts the efferent veins. Ejaculation is effected by contraction of the tubules of the epididymis, and of the vas deferens, followed, when excitement is intense, by a forcible contraction of the bulbocavernosus muscle of the urethra.

Female. The ovarian supply is similar to that of the testicle, while uterus and vagina are connected to the pelvic plexus. The autonomic does not play as important a part in sexual activity in the female; ovary and uterus are under hormonal control.

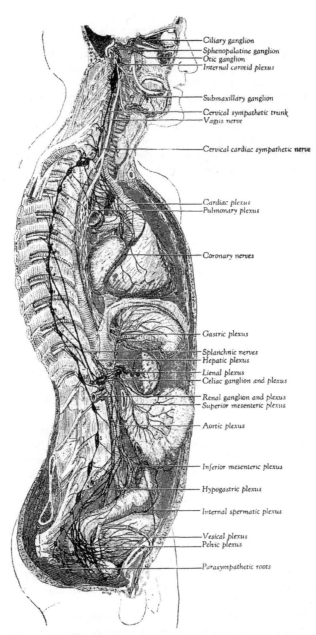

Ciliary ganglion
Sphenopalatine ganglion
Otic ganglion
Internal carotid plexus

Submaxillary ganglion

Cervical sympathetic trunk
Vagus nerve

Cervical cardiac sympathetic nerve

Cardiac plexus
Pulmonary plexus

Coronary nerves

Gastric plexus

Splanchnic nerves
Hepatic plexus

Lienal plexus
Celiac ganglion and plexus

Renal ganglion and plexus
Superior mesenteric plexus

Aortic plexus

Inferior mesenteric plexus

Hypogastric plexus

Internal spermatic plexus

Vesical plexus
Pelvic plexus

Parasympathetic roots

Figure 47. DISSECTION TO SHOW AUTONOMIC SYSTEM

The sympathetic system is rendered in solid black. The other nerves, including the parasympathetic, are white. The arteries are hatched.

(After Hirschfeld, from Krieg: Functional Neuroanatomy, McGraw-Hill Book Co., N.Y., 2nd edition, 1953)

ENDOCRINE ORGANS AND SPLEEN. The adrenal medullary cells present a postganglionic sympathetic ganglion whose cells have been transvested from conducting neurons into cells secreting adrenalin. Adrenalin is discharged into the blood stream during excitement or stress and enhances the effects of sympathetic stimulation when it reaches the postganglionic terminals.

The posterior lobe of the pituitary is under control of the hypothalamus. It produces a substance constantly which induces resorption in the kidney tubule of the excess of glomerular filtrate. Other endocrine glands are not primarily under autonomic control.

The spleen is supplied through the celiac ganglion and vagus. The smooth muscle in its capsule contracts when the sympathetic is active in excitement, causing the red blood corpuscles stored in its sinuses to enter the blood stream.

EYE. The ciliary muscle and muscles of the iris are under closer autonomic control than any other structures of the body. Their sympathetic supply is derived from the superior cervical ganglion, runs along the internal carotid artery, and passes through the ciliary ganglion without synapsing. The preganglionic axon leaves the cord at T1, and is under control of the postulated ciliospinal tract, which descends the brain stem from the midbrain. Sympathetic stimulation contracts the dilator pupillae, widening the pupillary aperture. In animals, more than in man, it protrudes the eyeball and widens the palpebral aperture. Because of the long course of the fibers, they are subject to injury, producing Horner's syndrome, characterized by constriction of the pupil, sinking of the eyeball, and partial closing of the eyelid. The parasympathetic supply of the eyeball has been discussed long ago (p. 50).

VASCULAR SYSTEM. The sympathetic fibers to the vessels to the thoracic and abdominal viscera come from the extensive plexuses which accompany them, but those to the limb arteries have a different routing. Those for the upper limb arteries arise and pass out of the spinal cord levels of T3-T7, synapse in paravertebral ganglia at these levels, ascend the sympathetic trunk and enter the roots of the brachial plexus (C5-T1) as grey rami. Those for the lower limb arise at cord levels T10-L2, run down the sympathetic trunk as postganglionic fibers, and enter the lumbo-sacral roots (L3-S3). The limb nerves send off twigs here and there which accompany arteries outward, branching and supplying their muscle. The extent of nerve supply parallels the relative development of smooth muscle, hence arterioles have more than large arteries, and large arteries more than veins. In only a few locations is there a parasympathetic supply.

The sympathetic contracts the blood vessels from cold to reduce heat loss, or in excitement, as though to reduce hemorrhage from wounds in combat. In Raynaud's disease the vessels are abnormally contracted, so, to treat it, the autonomic nerve supply to the affected limbs is interrupted at operation.

The sweat glands and hairs have only a sympathetic supply. Sweating is induced by its stimulation, but, paradoxically, sweat glands do not respond to adrenalin. The arrectores pilorum muscles on the hairs raise the hair shaft in excitement or in response to cold, making the familiar "goose-pimples."

CENTRAL CONTROL OF THE AUTONOMIC SYSTEM: HYPOTHALAMUS. Adjustment of visceral activity can be mediated by reflexes, completed at the cord or brain stem level, but generalized autonomic tone and autonomic reactions to emotions must be effected at a supraseg-mental level. This has proven to be the hypothalamus, the ventral part of the thalamic or diencephalic segment of the brain. It may be defined as the small region between anterior commissure (68: I), optic chiasma (77: I), and midbrain tegmentum behind. It is split into halves by the third ventricle (72: I). Laterally, it is bounded by the optic tracts (frontispiece). Although small, it is composed of about 20 nuclei, which as individuals are of tertiary importance because each has not been associated with a specific function. They will be enumerated briefly; reference is made to II and figure 39 and the tracing on p. 149 for identification. Beginning above and rostrally is the preoptic (139a), transitional between cerebrum and hypothalamus; the magnocellularis (139b), near the ventricle, supposed to conduct to the supra-opticus; filiformis (c); ventromedialis (d), central and ovoid and known to mediate the rage response; anterior (e); suprachiasmaticus (f), very small but quite specific; arcuatus (g), forming the ventral prominence of the tuber cinereum and continuing into the infundibular stalk of the pituitary; pre-mammillary (h), larger in lower mammals; mammillary (109: I L), described previously; tuberomammillary (j), laterally placed, its large cells and surrounding fibers indicating a motor nature; dorsomedialis (k), probably concerned with temperature regulation; posterior (m), transitional to the tegmentum; preopticus magnocellularis (n), belonging to the lateral or motor realm of the hypothalamus; supraopticus (o), along the optic tract, packed with large, very dark cells, and very vascular, and sending out the supraopticohypophyseal tract (q), which regulates secretion of the diuresis-inhibiting hormone of the posterior lobe of the pituitary; the lateral (p), probably the source of effector fibers; and the tuberis lateralis (r), peculiar to the higher primates and man.

The older and more obvious connections of the hypothalamus are in the olfactory realm (medial forebrain bundle, stria terminalis, stria medullaris, fornix). From perhaps having been a region where olfactory impulses influenced the viscera, the hypothalamus has become a region where all sorts of influences unite to control the autonomic system. One distinct tract is the periventricular tract which gathers from the hypothalamic nuclei, runs alongside the ventricle to the aqueductal region, down the brain stem as the dorsal longitudinal fasciculus. It ends largely in the periaqueductal grey matter and dorsal motor nucleus of the vagus. (See rat II M, Pv. T., D.L.F.)

Experimentation has shown that the hypothalamus has to do with a wide variety of generalized states and activities. They are listed below:

1. Temperature regulation. The posterior part increases heat production and prevents heat loss; the anterior part increases heat loss by vasodilatation and sweating.

2. Regulation of body fluid by the neurohypophysis has been mentioned.

3. Regulation of sugar metabolism. Certain hypothalamic lesions increase or decrease blood sugar, and response to insulin.

4. Regulation of fat metabolism. Basal hypothalamic operations are followed by storage of enormous amounts of fat in animals.

5. Gastrointestinal control. Lesions cause alterations in peristalsis or hemorrhage into the stomach.

6. Emotional expression. Severance of the forward connections or stimulation of the ventromedial nucleus produce uncontrolled outbursts of rage or flight.

7. Endocrine activity. In addition to a close control of the posterior pituitary, the adrenal cortex and medulla, thyroid, ovary, and some activities of the anterior pituitary, are affected by the hypothalamus.

It is generally accepted that the sympathetic center is in the posterior part of the hypothalamus, while the parasympathetic control is vested in its anterior region.

While the hypothalamus is the center for control of the autonomic system, many visceral effects are concomitants of cortical activity. Stimulation of the premotor area can induce changes in peristaltic activity, sweat secretion, hair erection, probably as an element of more general motor organization. The orbital areas 11 and 12 are, perhaps, more general areas of autonomic control which, in turn, preside over the hypothalamus, modifying autonomic control and saving us from being swept by devastating storms of visceral reflex.

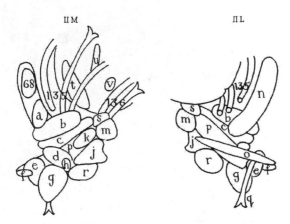

II M   II L

KEY TO DESIGNATIONS OF HYPOTHALAMIC NUCLEI AS REPRESENTED IN DIACHROME

(See also fig. 39)

a. Nucleus preopticus
b. Nucleus magnocellularis hypothalami
c. Nucleus filiformis hypothalami
d. Nucleus ventromedialis hypothalami
e. Nucleus anterior hypothalami
f. Nucleus suprachiasmaticus
g. Nucleus arcuatus hypothalami
h. Nucleus premammillaris
j. Nucleus tuberomammillaris
k. Nucleus dorsomedialis hypothalami
m. Nucleus posterior hypothalami
n. Nucleus preopticus magnocellularis
o. Nucleus supraopticus
p. Nucleus lateralis hypothalami
q. Supraoptico-hypophyseal tract
r. Nucleus tuberis lateralis
s. Zona incerta
t. Nucleus reuniens
u. Centrum medianum
v. Nucleus submedius
135. Ansa lenticularis
136. Fasciculus thalamicus

The following four illustrations form a continuous reconstruction of the nuclei and fiber tracts of the human brain stem. The numbers used are identical with those of the diachrome, and are indexed there and on pages 158-9.

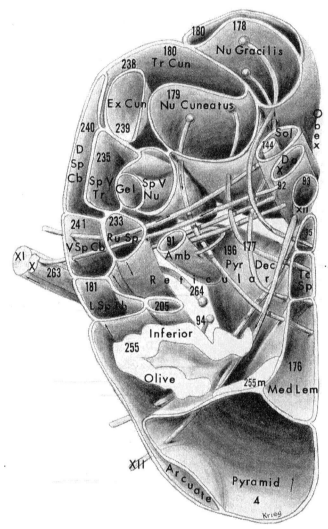

Figure 48. THICK SLICE RECONSTRUCTION OF LOWER
PART OF BULB, LOOKING CAUDALLY

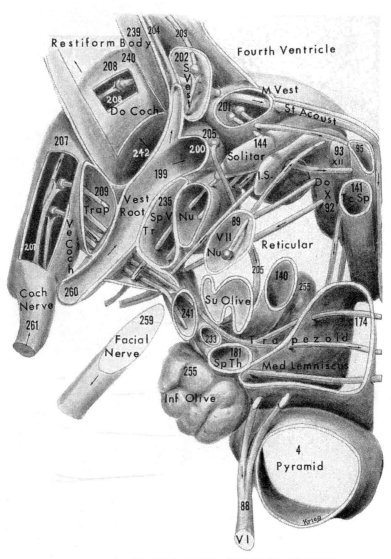

Figure 49. THICK SLICE RECONSTRUCTION OF UPPER
PART OF BULB, LOOKING CAUDALLY

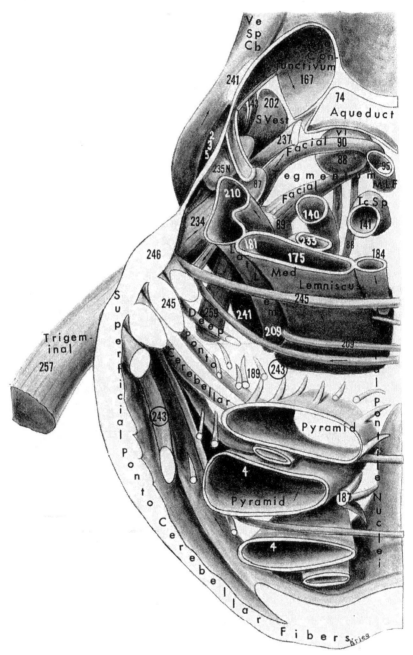

Figure 50. THICK SLICE RECONSTRUCTION OF PONS,
LOOKING CAUDALLY

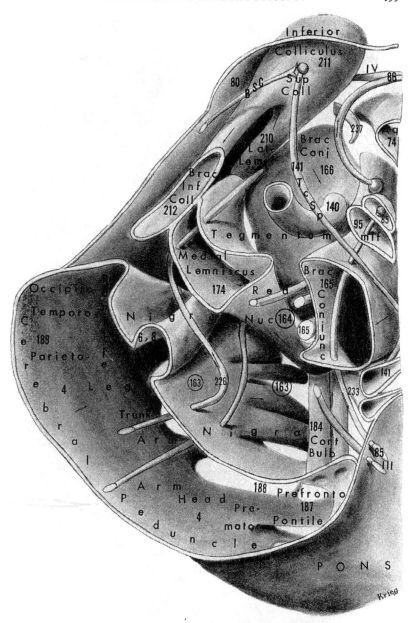

Figure 51. THICK SLICE RECONSTRUCTION OF MIDBRAIN
AND UPPER PART OF PONS, LOOKING CAUDALLY

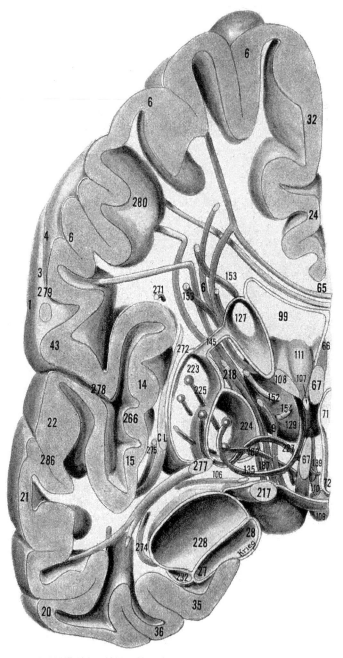

Figure 52. RECONSTRUCTION OF FRONTAL SLICE OF CEREBRUM,
INCLUDING STRIATUM AND ANTERIOR PART OF THALAMUS

The numerical references are indexed on pages 158-9, and correspond with the numbers of the diachrome.

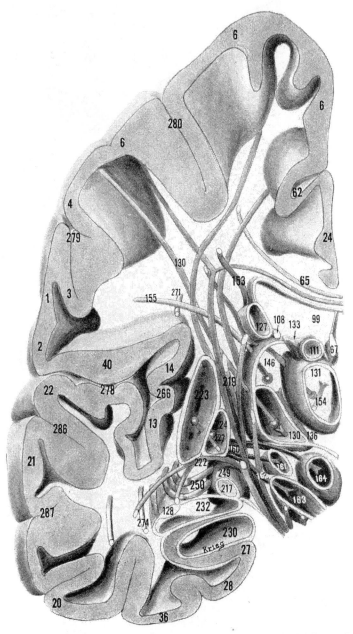

Figure 53. RECONSTRUCTION OF FRONTAL SLICE OF CEREBRUM,
INCLUDING MIDDLE PART OF THALAMUS

The numerical references are indexed on pages 158-9, and correspond with the numbers of the diachrome.

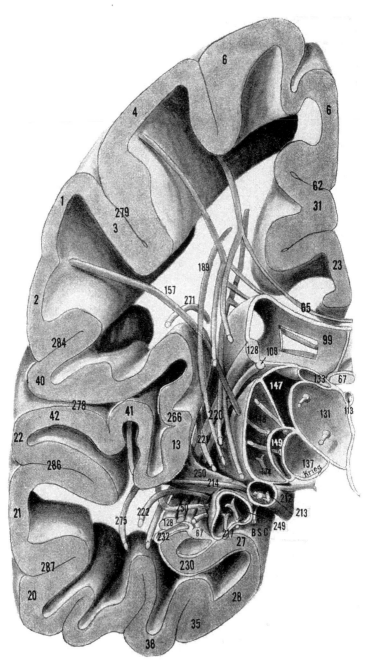

Figure 54. RECONSTRUCTION OF FRONTAL SLICE OF CEREBRUM, INCLUDING POSTERIOR PART OF THALAMUS

The numerical references are indexed on pages 158-9, and correspond with the numbers of the diachrome.

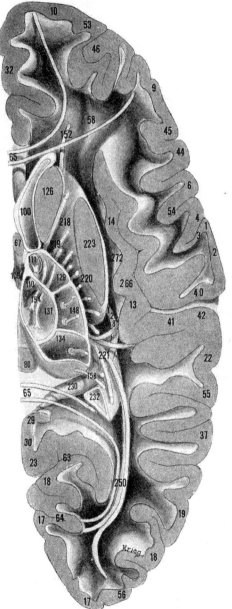

Figure 55. RECONSTRUCTON OF BASAL PART OF CEREBRUM

The numerical references are indexed on pages 158-9, and correspond with the numbers of the diachrome.

INDEX TO TEXT FIGURES AND TO DIACHROME

SURVEY OF THE DIACHROME
RECONSTRUCTION OF THE
MECHANISMS OF THE HUMAN BRAIN

The reconstruction of the human half-brain in medial and lateral aspects in diachrome permits representation of virtually every known significant cellular group and connection.

Each functional unit or system is given a different color. The three main sensory systems — somesthetic, visual and auditory — are in differing tones of blue. The motor systems are in tones of red. Cerebral cortex is basically yellow, but the yellow is locally blended with or replaced by the color of whatever functional system dominates any specific area, yielding various combined colors. The cerebellar systems are rendered in tones of brown; the olfactory system utilizes the purples.

The six double sheets of the diachrome may be thought of as a dissectible model of the right side of the brain. The model is like a box; sheet I, the medial surface, is the lid; sheet VI is the box itself, formed by the greater part of the cerebral cortex and its associational connections. The outer surface of the cerebellum and brain stem are on sheet V. Sheets II-V represent the internal nuclei and fiber tracts. II represents callosum, caudate and diencephalic nuclei; III shows thalamic radiations, somesthetic system, and the medial part of the cerebellum with the brachium conjunctivum; IV includes the entire cortical efferent system, with the addition of auditory and vestibular structures; while V presents lenticular, optic radiation and the remainder of the cerebellum.

All structures are designated by one system of ordinary numbers, the key to which is found on the inside of the front and back gate-fold sheets. Numbers 1-52 are reserved for the cortical areas and their fibers, using the standard designations of Brodmann. The others are applied sequentially, so far as possible.

Sheet O, facing I, and sheet VII, backing VI, give the functional significance of the cortical areas and main nuclear masses.

In the following summarizing description, the sequence of sheets, not the logical or functional order, will be followed, except that the details of the medial cortical surface and the lateral cortical surface will be considered together at the end of this account.

SHEET I

This sheet corresponds to the medial aspect of the right half of a hemisected brain, except that the brain stem portion is regarded as transparent, and shows the principal motor nuclei of the cranial nerves. The extensive multicolored region above, containing numbers 1 to 33, is the medial surface of the cerebral hemisphere. The colored patches are the individual cortical areas, having differing structures and functions. The surface irregularities are the convolutions, composed of ridges called gyri and valleys called sulci. Only several are constant and significant. The cingulate sulcus (62) separates the cingulate region below (areas 23, 24, 31, 32) from the somato-motor realm above (areas 1-12). The parieto-occipital sulcus (63) is deep and is separated from the upper end of the cingulate sulcus by the precuneus (areas 7, 31). The calcarine sulcus (64), containing the greater part of the optic projection, is separated from the preceding by the cuneus. Below is the receding basal surface of the brain which curves around the brain stem and cerebellum and reappears in front as the tip of the temporal lobe (20, 38). The red and orange sector of the cerebrum is the frontal lobe; areas 17-19, behind, form the occipital lobe, with the parietal lobe between (7).

Below the cortex is the callosum (65), a massive lamina composed of nerve fibers connecting the opposite sides of the cerebral cortex. Below is the fornix (67), a long, arching fiber tract, separated from the callosum by a thin partition, the septum (66), itself a neural structure belonging to the olfactory system.

All of these structures belong to the cerebral hemispheres, derived from the paired cerebral vesicles of the embryo. The structures below this are unpaired, though symmetrical, and contain the ventricular system of intercommunicating cavities: the slit-like third ventricle (71, 72), with its bridge (x), the massa intermedia, communicates in front with the paired lateral ventricles of the cerebral hemispheres by the interventricular foramina (73). Its roof is formed by a tuft of blood vessels, its choroid plexus (70). This segment of the brain is the diencephalon. Below is the brain stem, composed of midbrain (78), pons (59) and bulb (60). Midbrain and pons contain the narrow cerebral aqueduct (74), and the bulb houses the broad fourth ventricle (75, see overleaf), and its choroid plexus (76).

The motor nuclei and emergent fibers of the cranial nerves are distributed along the brain stem. The oculomotor nucleus (85) is oblong and sends its axons ventrally, to emerge on the midbrain. It sup-plies the ocular muscles, with the exceptions noted below. Just behind it is the tiny trochlear nucleus (86). Its axons run backward around the aqueduct (see overleaf), cross, and course around the opposite side of the midbrain. It supplies the superior oblique muscle of the eye. In the lateral part of the pons is the masticator nucleus (87), which sends its fibers into the trigeminal nerve (see V L, red strand with 257). It supplies the jaw muscles. The facial nucleus (89), at the ponto-bulbar junction, sends separate fascicles dorsally which hook sharply to form the genu (90), then run ventrolaterally to emerge as the facial nerve (259: V L), which supplies the muscles of facial expression. Within the genu of the facial nerve is the abducent nucleus (88). It sends its fibers directly ventrally, forming the abducent nerve to the lateral rectus muscle of the eye.

In the bulb the nuclei are elongate. The hypoglossal nucleus (93) is most medial, and just below the fourth ventricle. It sends a series of fascicles ventrally to emerge (265) as the hypoglossal nerve to the muscles of the tongue. Just laterally is the dorsal motor nucleus of the vagus (92), sending its axons laterally to emerge (263: V L), above the olive in the glosso-pharyngeal and vagus nerves. This nucleus furnishes the parasympathetic supply to the smooth muscles and glands of the viscera of thorax and abdomen. The ambiguus nucleus (91), extremely tenuous, sends its fibers out in the same rootlets as the preceding, but they end in the striated muscles of the pharynx and larynx. In the lower end of the bulb begins the motor column of the spinal nerves (94). Some of these nuclei are colored bright red, they belong to the somatic motor system. Others are a dull pink. They form the branchial motor and visceral motor systems.

The long green tract (95) is the medial longitudinal fasciculus, representing a primitive system for distribution of tonus, receiving contributions from the vestibular nuclei and distributing to the eye-muscle nuclei.

Above the brain stem is the vermis of the cerebellum, branched like an arbor vitae. Its cortex is colored, its fibrous core, sparse in the midplane, is white. Its three conventional lobes are shown, nearly equal in extent in this plane. Anatomically considered, they form the anterior (82, pink), middle (83, yellow), and posterior (84, blue) lobes. Functionally they are, respectively, somatocerebellum, cerebrocerebellum, vestibulocerebellum.

In front of the cerebellum, forming the tectum, are the superior colliculus (80), the optic reflex center; and the inferior colliculus (81), mediating auditory reflexes.

On the lateral side of the sheet are shown the associational connections of the medial cortex (their description will be postponed); the ventricular system, and most of the rhinencephalon, of olfactory nature.

The lateral ventricle is the much-reduced cavity of the cerebral hemisphere, 100 is its anterior horn, 99 its body, 98 its posterior horn and 97 is the beginning of its inferior horn, cut off, but continued in V as 232. The remainder of the ventricular system is shown (71-75); 101 is the lateral process of the fourth ventricle (actually lateral to the brain stem).

The rhinencephalon is rendered in shades of purple. The more obviously olfactory parts are deep violet, beginning with the olfactory bulb (103), which receives the olfactory filaments below, and sends back the lateral olfactory stria (cut); and the medial olfactory stria (105), which runs up to the septum.

One system is formed by the stria medullaris (107), which arches over the thalamus to end in the habenula (113), which in turn sends down the habenulopeduncular tract (114) to the interpeduncular

nucleus (115). Another system is composed of the fornix (67) ending massively in the mammillary bodies (109). These send upward the conspicuous mammillothalamic tract (110) to the anterior nucleus of the thalamus (111). A true thalamic nucleus, it has a cortical projection (112) which arches broadly to join the cingulum (96), and is distributed to the cingulate cortex. The stria terminalis (108) also belongs to the rhinencephalon. 117-125 designate the individual folia of the cerebellar vermis. (See p. 131 for names.)

SHEET II

The radiating arrangement of fibers is the callosum. Its component parts are not numbered, because their differential coloring corresponds to that of the areas which they connect. Those whose upper ends show a cut surface connect the medial cortical areas, depicted in I; the remainder may be traced through IV toward the lateral cortical areas in VI. It will be noted that the cingulate region has no callosal interconnections. Those to the temporal lobe arch over the lateral ventricle (view II L) and form a lamina along its side, the tapetum (follow in IV). (The callosal fibers run over the lateral ventricle, rather than under it, as a lateral view of II on I would indicate.)

Below the callosum is the caudate nucleus. 126 is its head, 127 its body, 128 its tail, continued in V M as 229.

In the concavity of the caudate are the thirty nuclei of the thalamus. The medial nucleus (131) dominates the medial aspect. The thin laminar nuclei (132) surround it except medially. Medial to it is the small habenula (113) the most primitive nucleus of the thalamus. The orange nuclei in front are ventralis anterior (129) and ventralis ventralis (130), medial to which are several of the median group of small and primitive nuclei. Above the medial nucleus is the almond-shaped later-

-alis dorsalis (133), which has no cortical projection; behind is the massive pulvinar (134). Below the medialis, and lateral to it, showing through, is the central nucleus (137), also without cortical connections.

The lateral aspect shows the parts of the ventral and lateral groups. Ventralis anterior shows again (129). Below it is ventralis ventralis (130), and proceeding backward are ventralis posterior inferior (150), and ventralis posterior lateralis (148). Ventralis posterior medialis, mostly hidden by 148, protrudes a little below (149). Of the lateralis group, 146 is lateralis anterior, 147 lateralis posterior and 134 pulvinar.

The inferior thalamic peduncle, shown on the medial surface, connects thalamus and pallidus (224; V M). 135 is an ansa lenticularis connection to lateralis dorsalis, behind is a similar fiber to ventralis anterior. 136 is the thalamic fasciculus, depicting a fiber to and one from the central nucleus, and a connection to the ventrolateral group. Their connections with the pallidus are shown in V M (135, 227).

Below and inferior to the thalamus the nuclei of the hypothalamus are reconstructed (139), but are too numerous and insignificant to catalog here.

Several structures of the brain stem are added in sheet II. The superior colliculus (80) sends fibers ventrally which decussate (x) as the dorsal tegmental decussation, and continues down as the tectospinal tract (141), which comes to lie over the medial lemniscus, and continues into the spinal cord. The central tegmental bundle (140) is also shown.

The most medial of the internal nuclei of the cerebellum is the fastigial nucleus (142). It emits the hook bundle (143) which crosses and ends among the vestibular nuclei.

The orange column is the solitary nucleus (144). It receives fibers conducting visceral sensation and taste. Its efferent fibers, being diffuse, are not shown.

SHEET III

The somesthetic system, represented in purplish blue, may be traced first. The ascending primary sensory fibers form the posterior columns of the spinal cord (180). These enter the secondary nuclei gracilis (178) medially, and cuneatus (179) laterally, and synapse. The secondary fibers (177) curve around the motor nuclei, decussate and turn rostrally as the medial lemniscus, which runs toward the thalamus, medially placed in the bulb (176), flattened in the pons (175), far lateral in the midbrain (174). The spinothalamic tract (181), already secondary, ascends lateral to the medial lemniscus, but merges with it in the pons. The secondary trigeminal tract (not shown) also joins the medial lemniscus, and thus augmented to include all of the ascending somesthetic fibers, it is distributed in the ventralis lateralis nucleus. The tertiary fibers arising here form the somesthetic radiation (157) to the cerebral cortex. They may be followed in IV and to the cortical areas 3 and 1 (VI, I). Their distribution is arranged systematically (O, VII), with the bodily segments in inverted order.

Other thalamic nuclei send radiations to the cortex, but little is known about what fibers supply them. Lateralis anterior and posterior project (155, 156) to the more posterior parietal areas. The posterior thalamic radiation from the pulvinar (158) runs back to parietal and occipital areas, while the part to the temporal lobe, including 159, the thalamotemporal fasciculus, curves, curves downward (trace in IV). Further forward are dorsally directed projections of the ventralis ventralis (130) and ventralis anterior to motor and premotor regions. All the dorsally running fibers compose the intermediate thalamic radiation. The anterior thalamic radiation (152-154, yellow) serves the medial nucleus and the prefrontal cortex.

It lies medial to the preceding, has a marked horizontal course, the more ascending fibers being crowded down by the callosum (153). All these thalamic radiations contain corticothalamic and thalamocortical fibers, but generally in inverse proportions.

The medial part of the lateral lobe of the cerebellum is shown, sending its axons centrally (172) to the internal cerebellar nuclei, the dentate (168) and emboliform (169). These, in turn, send forward the massive brachium conjunctivum (167), which crosses (166) in the midbrain, and continues forward (165) to the red nucleus (164) where many of its fibers terminate. Others continue to the nucleus ventralis ventralis, whose radiation ends in motor and premotor cortex.

The subthalamic group of nuclei are the nigra (163) and the subthalamic nucleus (161). They are served mainly by descending connections of pallidal origin, 162 and 160, respectively.

SHEET IV

All of the fibers which connect the cerebral cortex to other regions converge into one flattened bundle, the internal capsule (218-220). The fibers which are destined to leave the cerebral hemisphere converge toward its center. Near the surface they are mixed with associational fibers, but more centrally, where the fibrous cores of the gyri fuse, they form a continuous mass, the corona. In the drawing, where fibers are separate, they are within their gyral cores. Laterally, where obstructed by massive formations, they run upward before turning downward (44-46), and in the temporal lobe they run backward, then medially, before joining the corona. The cingulate areas apparently contribute no fibers to this system, the central region supplies most, and as the frontal and temporal lobes are approached, the participation diminishes.

From the corona emanate the callosal fibers (II), leaving the thalamic and true projectional components to form the internal capsule. Medial to the capsule are caudate and thalamus, lateral is the lenticular nucleus (223, 224: V) compressing the bundle into a fan, while cellular strands between caudate and lenticular perforate it. The anterior limb (218) is the part lateral to the head of the caudate, the posterior limb (219, 220) is the massive part lateral to the thalamus. Continuing backward, the part to the occipital lobe is the retrolenticular capsule (221). The contingent to the temporal lobe, running laterally, is the infralenticular capsule (222). A large part of the capsule is formed of thalamic connections which gradually blend into it (match III with IV).

At the lower end of the thalamus the cerebral peduncle begins as a massive bundle on the ventral aspect of the midbrain, composed exclusively of cortical projection fibers. From area 6, fibers diverge to end in the midbrain (182) and subthalamus (183), while the medial corticobulbar tract traverses the medial lemniscus to terminate in the tegmentum of the brain stem (184-186).

In the pons the peduncle separates into fascicles. Among its cells the contingents from most of the areas terminate. The frontopontile group (187) stems from areas 8-12. The premotor group, from 6, are numerous (188). There are many fibers from the most posterior parietal areas (189), and a few from the occipital lobe.

Virtually the only projection fibers to run the gamut are the pyramidal system (red) from area 4. As the pyramids, they continue on the ventral surface of the bulb, and at its caudal end cross in large bundles, the decussation of the pyramids (196) to enter the lateral funiculus of the spinal cord as the lateral corticospinal tract (197). A few remain uncrossed and continue in the cord as the ventral corticospinal tract (198). Along the way, in the brain stem, small contingents of fibers separate from the pyramidal tract to supply the cranial motor nuclei. These are the corticobulbar fibers and include components to masticator and facial nuclei (190, 191), to motor nuclei in the bulb (192-194) and the recurrent Pick's bundle (195).

On sheet IV are also reconstructed the vestibular system in bluish green and the auditory system in dark blue. Both enter by the auditory nerve (199, 206). The vestibular nerve immediately branches into ascending and descending roots, which distribute the secondary vestibular nuclei. Some continue to the cerebellum. The secondary nuclei are: superior (202), medial (201), spinal (200), and lateral, amidst these three. The lateral nucleus gives origin to the vestibulospinal tract (205). The others send most of their fibers to the medial longitudinal fasciculus (95: I), but the superior sends to the cerebellum (203, 204).

The cochlear nerve (206) breaks up in the ventral and dorsal cochlear nuclei (207, 208), plastered onto the shoulder of the bulbopontine junction. From the dorsal nucleus, fibers run medially on the upper surface of the bulb (thin blue strand) as the auditory striae. From the ventral nucleus emanates the more considerable trapezoid body (209) which crosses the midplane and turns rostrally as the lateral lemniscus (210). The lateral lemniscus ascends the midbrain along its lateral surface, and on reaching the inferior colliculus (211), some of its fibers enter and form reflexes, others continue as the brachium of the inferior colliculus (212) to the medial geniculate body (213), a thalamic nucleus. Synapsing here, the auditory fibers, now tertiary, form the auditory radiation (214), which has only a short way to reach the auditory cortex (41: VI).

The optic nerve (215), chiasma (216)

and tract (217) are shown curving around the cerebral peduncle on their way to the lateral geniculate body.

SHEET V

The broad blue lamina represents the continuation of the optic system. The optic tract ends in the lateral geniculate body (249), helmet-shaped, an outlying thalamic nucleus. From it emerges the large optic radiation (250). The upper part curves around the concavity of the lateral ventricle and proceeds backward directly, the lower part (251) is pulled forward into the temporal lobe by the inferior horn of the lateral ventricle. Alongside the ventricle and its posterior horn, the optic radiation forms a vertical band, the external sagittal lamina, from the edges of which fibers peel off to pass to their termination along the calcarine sulcus (17: I), at first those from the peripheral monocular fields (252), then the binocular fields (253); while the macular fibers (254) proceed to the occipital pole.

The lenticular nucleus belongs, with the caudate, to the striatal complex. It is composed of putamen (223) laterally, and pallidus (224) medially. Caudate and putamen are connected by cellular strands (145), through which pass fibers. The caudate and putamen contain pencils of fibers (225) converging into the pallidus, where they end. The pallidus, likewise, contains numerous fibers which converge toward its lower surface and medial edge, and leave to form the following fiber groups which terminate in other structures. Ansa lenticularis (135) is composed of ventrally directed fibers in the pallidus which leave its ventral surface, pass medially and loop under the lowest part of the anterior limb of the internal capsule. They enter ventralis anterior and lateralis dorsalis nuclei of the thalamus (II). The fasciculus thalamicus (136: II) and pallido-bulbar and pallido-olivary bundles (226) jog dorsally between pallidus and capsule, then thread through, to end where their names indicate. The strionigral group is radial within the pallidus, leaves at its apex, threads through the capsule, but at a lower level, and ends in the nigra (III: 162).

The inferior horn of the lateral ventricle is shown here (232) because it is so far lateral. Along with it runs the tail of the caudate nucleus (229), which joins the base of the putamen; and the stria terminalis (108), the chief tract of the amygdala (228), a large nucleus, of intermediate nature between striatum and cortex. The hippocampus, a primitive type of cortex (230), curves along the ventricular horn and ends in an expanded foot (231). It gives rise to the fornix (67: I).

This sheet shows the connections of the cerebellum, except those of vestibular origin. The most massive of the three cerebellar peduncles is the middle, or brachium pontis arising from the pontile nuclei (243) of the opposite side. The pontocerebellar fibers accumulate superficially (244) and deeply (245) of the pontile nuclei, converge laterally to the brachium pontis (246) and enter the lateral lobes of the cerebellum, to form the bulk of its medullary center. Within the folia the medullary center forms laminae (247), which distribute their fibers to the continuous folded mantle of cerebellar cortex (248). The somatic cerebellar afferents enter through the inferior peduncle, or restiform body (242). The following contributions are shown. The spinocerebellar fibers from the upper limb and neck (238) synapse in the external cuneate nucleus (239), and are relayed into the restiform body. The dorsal spinocerebellar tract (240) veers gradually into the restiform body, but the ventral spinocerebellar tract keeps its rostral course and turns back (241) on the brachium conjunctivum, to join the restiform body within the cerebellum. The inferior olive is a shell-like nucleus (255) in the ventral part of the upper bulb, so large that it

forms a surface prominence (see overleaf). It receives the strio-olivary tract (226) and sends out of its medial surface many fibers which cross, turn dorsally and join the restiform body as the olivocerebellar fibers. Within the cerebellum the restiform body curves over the lateral and dorsal surfaces of the dentate nucleus, and is distributed chiefly to the vermis region, particularly of the anterior lobe (170: III).

The trigeminal nerve and nuclei are colored greenish blue. On the lateral surface the three branches, ophthalmic, maxillary and mandibular, are shown, joining at the semilunar ganglion (258), where the primary cell bodies are located. The large sensory root (257) enters the brachium pontis far laterally, and may be seen on the medial aspect (234) proceeding dorsomedially. It turns caudally and continues down to the upper end of the spinal cord as the spinal tract of the trigeminal (235). Along its medial surface lie the spinal nuclei of the trigeminal, which contain the secondary cell bodies for the reflex and projectional trigeminal pathways (not shown). Some fibers synapse in the globular main sensory nucleus (236) placed at the junction of the trigeminal root and spinal V tract. The mesencephalic root (237), proprioceptive for mastication, turns rostrally and runs along the aqueduct where, paradoxically, its primary cells are located.

The rubrospinal tract (233), crossing above the ventral tegmental decussation (x), is shown.

The lateral aspect of sheet V shows the outer surface of cerebellum and brain stem. The folia of the lateral lobes (248) form a series of ridges ending at the transverse fissure. Just in front is the flocculus (256), the vestibular part of the lateral lobe. The pons is perforated by the trigeminal (257) and facial (259) nerves. At the cerebello-bulbo-pontile angle are the vestibular (260) and cochlear (261) divisions of the acoustic nerve. Over the top of the olive (255) emerge the row of filaments that compose the glossopharyngeal (262) and vagus (263) nerves. The accessory nerve (264) ascends along the bulb and joins this group. Below the olive is the hypoglossal nerve (265).

SHEET VI

In the last sheet of the series the cerebral cortex is regarded as a shell, the medullary center being considered hollow, except where sample associational connections are reconstructed. The surface is deeply furrowed by the sulci, which show as elevations on the inner aspect, but the most extensive configuration is the insula (266), easily visualized from the inner aspect. It is an expansion of the base of the lateral fissure and is closely related to the lenticular. Between the two is a thin fiber lamina, the external capsule (272, 275). Actually, a thin cellular sheet, the claustrum (not shown) intervenes between external capsule and insula, and a further thin fiber lamina intervenes between claustrum and insula, the extreme capsule. This lamina is composed of fibers serving the insula, while the external capsule is composed of long associational fibers.

The longest associational fibers run from occipital to frontal poles, and vice versa, the occipitofrontal fasciculus (275). Its lower fibers are curved at the middle and connect the temporal and orbital cortex, the uncinate fasciculus (276). More dorsally placed longitudinal fibers are pushed into an arch by the insula and form the arcuate fasciculus (271). The anterior commissure (277) is a special callosal commissure for the middle temporal gyrus. It is a landmark in the hemisected brain (68: I) and marks the rostral extreme of the unpaired portion of the brain. On the basal surface is the extensive inferior longitudinal fasciculus (274) connecting occipital and temporal regions. External to these systems, in the posterior half of the cerebrum is a very extensive lamina of fibers whose general course is obliquely downward and forward, connecting parietal to temporal stations. This is the extreme sagittal stratum (273). Further outside and also above it are extensive parietal associational systems which are less simply organized. All these may be seen in fiber dissections of properly treated brains, but our knowledge of shorter internal connections is derived from Marchi degeneration studies of primates and from oscilloscopic studies. The more important of them are shown by arcuate fibers in the upper part of VI M and I L. They carry the colors of their area of origin.

The sulci of the lateral surface of the cerebrum are shown in VI L. The lateral fissure (278), very deep, separates the temporal lobe from frontal and parietal lobes. The central sulcus (279) is important as separating motor from sensory cortex, and frontal from parietal lobes. The precentral gyrus intervenes between the central sulcus and the precentral sulcus (280). In the frontal lobe the superior (281) and inferior (282) frontal sulci delimit superior, middle and inferior frontal gyri. The general regions marked 47, 45, 44, respectively, correspond to orbital, triangular and subcentral opercula, which are billows or flaps, with cortex on both surfaces, covering the insula.

The vertical gyrus behind the central sulcus (3, 1, 2) is the postcentral. It is limited behind by the postcentral sulcus (284). The remainder of the parietal lobe is irregularly divided by the intraparietal sulcus (285) into superior and inferior parietal lobules. The temporal lobe is divided by the superior (286) and middle (287) temporal sulci into superior, middle and inferior temporal gyri (22, 21, 20, respectively).

We may now survey the cortical areas. Their functions, so far as known, are indicated on sheets VII for the lateral cortex and O for the medial cortex. Area 3, mostly in the central sulcus, receives the somesthetic projection. It connects with areas 4, 6, 1, 2, 5, 7 (see VI M and I L for all associations). Area 1 is similar but receives fewer somesthetic fibers. Area 2 receives from 3, 1, 4 and 6 and distributes widely to parietal areas 7, 39, 40, 1 and 3, to precentral areas 4 and 6, and prefrontally. It has extensive callosal connections. Areas 5 and 7 receive from 3, 1, 2 and in turn distribute to precentral areas 4 and 6, widely in the parietal lobe. They also send a component to the pons. Areas 39 and 40 associate extensively in the parietal sector. Through the extreme sagittal stratum they connect with the temporal region, and through the arcuate fasciculus with the motor speech area in inferior frontal gyrus.

In the occipital lobe, area 17 receives the visual projection and connects with 18. Area 18 distributes to 19 and 37. Area 19 connects widely—with 37, 39, 21, and projects to the pons.

The auditory area 41, receives the auditory projection and distributes to the neighboring 42 and 22. The temporal areas are poorly understood, but they receive from the associational areas 39, 40, 19, 37, and project mostly to the orbital, polar and opercular regions of the frontal lobe, through the uncinate and arcuate fasciculi.

The motor area 4 receives from a great variety of areas, parietal (3, 1, 2, 5, 7) and frontal (6, 8, 9) and sends down the pyramidal system (IV). It has only small callosal and thalamic connections. In general it sends to regions from which it receives. (This is true of cortical associations and thalamic projections generally.)

The premotor area 6 receives from the frontal areas ahead of it (8, 9), from 5 and 7, and from nucleus ventralis ventralis. It associates widely and with the opposite side. It distributes extensively to area 4, and through the capsule to subthalamic, tegmental and reticular motor nuclei, and to the pons. Area 8 receives from medius and submedius nuclei, from frontal areas ahead of it, and distributes to tegmental motor structures and eye muscles, also to 6. The prefrontal areas 9, 10, 46 receive from the medial nucleus, basal frontal and temporal areas, and send to 8, 6 and the pontile nuclei. The orbital frontal areas (11, 12, 16) receive from the medial nucleus and temporal lobe, and distribute to prefrontal areas and hypothalamus. The opercular areas 44, 45 receive from the parietal association areas 39, 40 and must distribute to the head region of the premotor and motor areas. Much remains to be discovered of cortical connections.

NOTES ON
THE RECONSTRUCTION OF THE RAT BRAIN

The reconstruction of the rat brain, in halftone on paper, and placed in the pocket in the back cover of this book, is planned to match the diachrome of the human brain, structure for structure. Similar parts are similarly numbered. There are, however, only eight pages, or four sheets, to the rat reconstruction. Sheets II and III of the diachrome are combined into a single sheet in the rat reconstruction; likewise sheets V and VI of the diachrome are combined into one.

Instead of separate colors, distinct tones of gray were used to represent the several functional systems, as far as this was possible. The structures in each sheet have been represented against the brain as a background, to simulate the appearance of the assembly of sheets in a corresponding opening of the diachrome.

In the following account only the divergencies from the human reconstruction will be signalized; the summary of the diachrome can be used for a general description of the rat reconstruction, when the differences here noted are taken into consideration.

SHEET I

The degree of correspondence between the brains of two forms as widely separated as man and rat is remarkable, as is shown when they are reconstructed as they are here. The cerebral hemispheres are not as well developed, so do not override the cerebellum. This permits the long axis of the cerebrum and brain stem to coincide.

The cingulate areas (23, 24) occupy nearly all of the medial surface of the cerebral cortex; the retrosplenial area (29) is especially large in the rat, and must be important, inasmuch as fibers converge there both from cingulum and from neocortical areas.

Advantage is taken of the large size of the massa intermedia to depict some of the midline group of thalamic nuclei, which are relatively large in the rat. They are: Pt., paratenialis; Pv. St., paraventricularis stellatocellularis; Pv. Ro., paraventricularis rotundocellularis; Rh., rhomboidalis; Re., Reuniens; Sm., submedius (compare fig. 38, p. 75).

The plans of the cerebellar vermis, and of the motor cranial nerve nuclei, are so similar in rat and man that they do not require commentary.

On the lateral side of the sheet the configuration of the lateral ventricle has been omitted, since its posterior part is broad and vertical, with the inferior horn poorly developed, and important features of the reconstruction would be covered.

The structures of the rhinencephalon are much better developed in the rat than in man. The olfactory bulb (103) is large, and nearly terminal. In lower forms (see figs. 11-13, pp. 30-32) it is quite terminal. The olfactory glomeruli, formed by synapse of olfactory nerve fibers with dendrites of mitral neurons of the olfactory bulb (fig. 1, H), occupy the outer surface of the bulb. The olfactory tract (104) carries the impulses backward and trifurcates into medial (105), intermediate (I. S.) and lateral (L. S.) striae. The medial stria passes to the large septal nuclei (66), which in turn send the diagonal band (106) to the basal olfactory region. The intermediate stria ends in the olfactory tubercle (T. O.), so small in man that it could not be represented. The broad lateral olfactory stria runs nearly the length of the ventral surface of the cerebrum (51: VI L). In the rat the main portion of the anterior commissure (68) connects the olfactory bulbs; in man this component is suppressed.

Only relative sizes and proportions differentiate the other rhinencephalic structures of rat and man; there is no further differentiation in man. Stria medullaris (107), habenula (113), habenulopeduncular tract (114), interpeduncular nucleus (115), stria terminalis (108), fornix (67), mammillary body (109), mammillothalamic tract (110) and anterior nucleus of thalamus (111) are better developed in the rat. All this is correlated with the highly olfactory sensory life of the rat.

SHEET II - III

In this sheet the thalamic radiations have been combined with the representation of the thalamic nuclei; and the few structures of the brain stem shown in human sheet II have been combined with the more numerous ones of sheet III.

The morphology and fiber arrangement of the callosum (65) are similar to man's. In the rat, however, we have more detailed data on the contribution made by each area, and we find they differ widely. The frontal area 10 has a rich interhemispheric connection, while motor area 4 has none. The somesthetic area (2) and the auditory area (41) are well represented, but not the visual area (17). The entorhinal (28) and perirhinal (35) areas, also area 20, are well represented, the other association areas less so, and the cingulate region apparently not at all.

One considerable difference in the two brains is that caudate and putamen are not separated. The anterior part of the internal capsule thus is not discrete, and the forward part of the striatum is shot full of fiber bundles belonging to the capsular

system and the thalamic radiation. The blocky caudate (126) is cut laterally to indicate the arbitrary separation from the putamen.

The thalamus is relatively small, and except for an elaborate set of "midline" nuclei, has a simpler nuclear pattern than man. There is a medial nucleus (131), which projects forward, through the anterior thalamic radiation (152), to the frontal areas 10 and 8. V. M. is the large ventralis medialis, Pf., the parafascicularis. These connect to the basal frontal region. 132 is centralis lateralis, one of the intralaminar group. Laterally, the undifferentiated ventral or somesthetic nucleus (148) and the lateral nucleus (146) are visible. From them emanate the intermediate thalamic radiation, more compact than in man, because of the guillotining effect of the hippocampus (V: 230), fornix (67) and stria terminalis (108). Although the anterior radiation is considerable, the greatest component is the somesthetic projection system (157). The ascending somesthetic pathway (174-180) is much the same as in man, and may be traced in this sheet. It is represented in dark gray. The radiation of the lateral nucleus (155) is weaker, and passes to the more posterior non-sensory cortical areas (VI L: 18, 18a, 37, 39, 40).

The hypothalamus is relatively larger and lies directly under the thalamus. Its nuclei correspond in large part to the human hypothalamic nuclei. They may be identified and compared by reference to the diagram and key on page 149. It emits the periventricular tract (pv. t.), which continues down the brain stem as the dorsal longitudinal fasciculus (D. L. F.). This tract may be concerned with sleeping and waking, or with visceral reflexes. It is present in man also, but not depicted.

SHEET IV

As in the diachrome, this represents the

cortical projection system, vestibular system and cochlear system. The shading of the cortical projection neurons corresponds to that of their areas of origin, the vestibular system is very light, the cochlear system is middle gray. The optic tract (217) is also represented.

In the rat nearly every area sends some contribution to the capsular projection system, but a large proportion leave at the midbrain level. There is a considerable projection from the prefrontal-premotor region (10). Since this area receives the entire projection from the medial nucleus, it can hardly be regarded as motor or purely premotor, but its large and perseverating projection forestalls its being regarded as purely prefrontal. The fibers remain medial in the peduncle and continue through the bulbar pyramid. The motor area (4) and premotor area (6) send many fibers down from the peduncle and pyramid into the cord. There is a cortico-reticular component from area 6, as in man.

The somesthetic area (2, dark gray) apparently represents 3, 1, 2 and 5 of the human brain, for it not only sends many fibers back into the ventral nucleus (IV M) but also into the tegmentum, pontile nuclei and, seemingly, to the nigra. The other sensory receptive areas give out descending fibers: from the visual receptive area (17) to the lateral geniculate, and from the auditory area (41) to the medial geniculate and inferior colliculus. As is apparently the case in man, division of labor has not proceeded as far in the auditory cortex as in the visual realm, since all or most of the very numerous descending connections to the *superior* colliculus arise in the surrounding area 18 and 18a.

All of the capsular connections of the posterior third of the cortex are forced far out of their way by the already-mentioned constricting influence of the stria terminalis, hippocampus, fornix and optic tract, making a large retrolenticular internal capsule. This is not quite the same as in man, for here the barrier is behind, while in man it is the putamen, in front. The number of retrolenticular fibers per unit of neocortical area is reduced, as compared with the more forward regions; likewise much of the posterior cortex is of the rhinencephalon, which does not contribute to the capsule. In the rat, area 29, at any rate, of the cingulate areas, sends long projection fibers into the capsule. The basal neocortex (13, 14) contributes fibers to the lateral extreme of the peduncle. Areas 2 and 6, and in smaller numbers other parietal regions, connect with the pontile nuclei. There is a typical ventral pyramid (194) with a decussation (196), but the crossed corticospinal tracts (197) run down the dorsal columns of the cord, rather than down the lateral columns. It is formed chiefly of fibers from 4, 6, 10 and 2, in that order of frequency.

The rat has acute hearing, and the cochlear system is well developed. The ventral cochlear nucleus (207) is the larger, forming the typical mammalian acoustic tubercle. The ventral nucleus (208) relays the discriminative fibers, and these are relatively fewer than in man. Acoustic striae (S. A.) are present, but the trapezoid body (209) is larger. Inferior colliculus (211) is huge, and, as in man, has no visible structure. Medial geniculate (213), of fair size, sends the auditory radiation (214) to the auditory cortex (41) on the lateral prominence of the cerebrum.

The vestibular system is conservative, and shows the same nuclei as man, except that the lateral vestibular nucleus (L.) is relatively larger than in man.

SHEET V - VI

Because there are few association pathways in the rat, sheets V and VI can be combined without great loss. They form a thin sheet just under the cortex and in its deeper layers, and for the most part, follow an oblique medio-caudal direction. They are located above and outside the other structures. A considerable group run from area 2 to area 7, and from 18 to 17, but there is a marked tendency to continue medially to the edge of the cingulum and then to sweep to more caudal stations on the cingulate and retrosplenial cortex. This cingulate component is known to exist in the monkey, and no doubt does in man. Areas 17, 18, and 18a interconnect abundantly, following this oblique direction of the other associational fibers for the most part. Area 2 also sends forward to the frontal region, and back to 40. Auditory cortex (41) also connects to 18 and 18a, along with visual area 17, forming, with 7, a common sensory association cortex. There is evidence that the basal olfactory cortex (51) contributes to the oblique association lamina, forming a paleo-neocortical connection.

The large, blocky structure is the putamen (223), medial to which is the pallidus (224). There is a rich and varied pallidal outflow comprising an ansa lenticularis (135), a fasciculus thalamicus (136), and hypothalamicus. There are numerous descending pallidals (226) to subthalamus, tegmentum and olive.

The lack of pigment in the eye of the albino rat makes good vision impossible, hence the optic system is small, with a small lateral geniculate (249) and a quite small optic projection (250) curving around the lateral ventricle, to area 17.

The hippocampus (230) is very large, and shaped like a curved Parker House roll, forming the most forward part of the recurved medial posterior cortex. It has not migrated as far away from its pristine forward position as in man. The dentate gyrus (D.) is behind, the hippocampus proper (H.) in front. The fornix (67) and hippocampal commissure (not shown) are large.

The amygdala is large and divisible into basal (228b), medial (228m) and cortical (228c) portions. The cortical areas peripheral to the hippocampus are well displayed in the rat. In order they are: 27, 49, and the entorhinal area 28; with 35 outside this, on the lateral surface.

Cerebellar connections are much the same as in man, but the small size of pontile nuclei (243) and brachium pontis (244) is interesting. Consequently, the cerebrocerebellum (83) is much smaller laterally than in man. The flocculus (256) is large.

The spinal trigeminal tract and nucleus are very large (235), correlated with its exquisite muzzle sensitivity, and the vibrissae.

Most of the cortical areas of the lateral aspect have been mentioned. The three sensory areas (3, 17, 41) are relatively large, and almost equal. The associational areas between and surrounding them are relatively small. They include areas 18 and 18a, respectively, medial and lateral to 17, chiefly optic, but partly auditory associative; area 7, parietal associative and projectional, and areas 39 and 40, somesthetic associative. These are all well developed, with numerous afferent and efferent fibers. Areas 37, 20, 14 and 13 are not well differentiated, and have fewer fibers.

Area 4 has position and connections which indicate it to be a counterpart of the human motor area, but it lacks giant pyramidal cells. A small strip just in front, called 6, has more the structure of a premotor area. The rather large area 10 + forming the frontal region has attributes of premotor and prefrontal cortex, but the structure of premotor cortex. Its stimulation leads to discrete movements, so must be regarded as functionally motor. Area "8" below, connected with the parafascicularis, is difficult to correlate; the frontal region of the rat is so slight, and that of man so vast.

BOOKS FOR FURTHER READING

TEXTS

BRODAL, ALF: Neurological anatomy in relation to clinical medicine. 496 pp. 1948. Oxford, England.

BUCHANAN, A. R.: Functional neuro-anatomy. 362 pp., 273 figs. 1957, 3rd ed. Lea and Febiger, Philadelphia.

ELLIOTT, H. CHANDLER: Textbook of the nervous system. 437 pp. 1954, 2nd ed. Lippincott, Philadelphia.

GARDNER, ERNEST: Fundamentals of neurology. 359 pp., 142 figs. 1952, 2nd ed. W. B. Saunders Co., Philadelphia.

KRIEG, WENDELL J. S.: Functional neuroanatomy. 659 pp., 326 figs. + 87 atlas pages. 1953, 2nd ed. McGraw-Hill Book Co., 330 W. 42nd St., New York.

KUNTZ, ALBERT: Textbook of neuroanatomy. 524 pp., 331 figs. 1950, 5th ed. Lea and Febiger, Philadelphia.

LARSELL, OLOF: Anatomy of the nervous system. 520 pp., 382 figs. 1951, 2nd ed. Appleton-Century-Crofts, New York.

METTLER, FRED A.: Neuroanatomy. 536 pp., 357 figs. 1948, 2nd ed. C. V. Mosby Co., St. Louis.

PEELE, TALMAGE L.: Neuroanatomical basis of clinical neurology. 564 pp., 313 figs. 1955. McGraw-Hill Book Co., 330 W. 42nd St., New York.

RANSON, STEPHEN W., and SAM L. CLARK: Anatomy of the nervous system. 581 pp., 434 figs. 1953, 9th ed. W. B. Saunders Co.

STRONG, OLIVER S., and ADOLPH ELWYN: Human neuroanatomy. 488 pp., 357 figs. 1953, 3rd ed. Williams and Wilkins Co., Baltimore.

ATLASES

BASSETT, DAVID L.: A stereoscopic atlas of human anatomy. Section I. 1952. Sawyer's, Inc., Portland, Oregon.

OLSZEWSKI, J., and DONALD BAXTER: Cytoarchitecture of the human brain stem. 199 pp., many figs. 1954. S. Karger, New York.

RASMUSSEN, ANDREW T.: Atlas of cross section anatomy of the brain. 63 figs. 1951. McGraw-Hill Book Co., 330 W. 42nd St., New York.

RILEY, HENRY A.: Atlas of the basal ganglia, brain stem and spinal cord. 708 pp., 258 figs. 1943. Williams and Wilkins Co., Baltimore.

SINGER, MARCUS, and PAUL I. YAKOVLEV: Human brain in sagittal section. 81 pp., 45 figs. 1955. Charles C Thomas, Springfield, Ill.

ADJUNCTS

McDONALD, JOSEPH, and JOSEPH G. CHUSID: Correlative neuroanatomy and functional neurology. 263 pp., 177 figs. 1952, 6th ed. University Medical Publishers, Los Altos, Calif.

HAUSMAN, LOUIS: Atlas of consecutive stages in the reconstruction of the nervous system. 81 pp., 70 figs. 1953. Charles C Thomas, Springfield, Ill.

NETTER, FRANK H.: Ciba collection of medical illustrations. I: Nervous system. 143 pp., 104 plates. 1953. Ciba, Summit, N.J.

RASMUSSEN, ANDREW T.: Principal nervous pathways. 73 pp., 28 figs. 1952, 4th ed. Macmillan Co., New York.

CORTEX

BAILEY, PERCIVAL, and GERHARDT von BONIN: Isocortex of man. 301 pp., 121 figs. + 15 plates. 1951. University of Illinois Press, Urbana.

BAILEY, PERCIVAL; GERHARDT von BONIN and WARREN S. McCULLOCH: Iso-cortex of the chimpanzee. 440 pp., 293 figs. + 40 plates. 1950. University of Illinois Press, Urbana.

KRIEG, WENDELL J. S.: Connections of the frontal cortex of the monkey. 299 pp., 187 figs. 1954. Charles C Thomas, Springfield, Ill.

von BONIN, GERHARDT: Essay on the cerebral cortex. 150 pp., 32 figs. Charles C Thomas, Springfield, Ill.

von ECONOMO, CONSTANTIN: Cytoarchitectonics of the human cerebral cortex. 186 pp., 61 figs. 1929. Oxford University Press, New York.

COMPARATIVE

JOHNSTON, J. B.: Nervous system of vertebrates. 370 pp., 180 figs. 1906. P. Blakiston's Son and Co., Philadelphia.

KAPPERS, C. V. A.; G. C. HUBER and E. C. CROSBY: Comparative anatomy of the nervous system of vertebrates, including man. 1845 pp., 710 figs., 2 v. 1936. MacMillan Co., New York.

MEESEN, H., and J. OLSZEWSKI: Cytoarchitectonic atlas of the rhombencephalon of the rabbit. 52 pp., 68 figs. 1949. S. Karger, New York.

OLSZEWSKI, JERZY: Thalamus of Macaca mullata. 93 pp., 60 figs. 1952. S. Karger, New York.

PAPEZ, JAMES W.: Comparative neurology. 518 pp., 315 figs. 1929. Thos. Y. Crowell Co., New York.

CLASSICS

CAJAL, S. RAMON Y: Histologie du systeme nerveux. 986 + 993 pp., 443 + 582 figs., 2 v., 1909, 1911; reprinted 1955. Consejo Superior de Investigaciones Cientificas. Madrid.

DEJERINE, JULES: Anatomie des centres nerveux. 816 + 720 pp., 401 + 465 figs., 2 v., 1895, 1901. J. Rueff, Paris.

WINKLER, CORNELIUS: Anatomie du systeme nerveux. 5 v., 1921. De Erven F. Bohn, Haarlem.

PHYSIOLOGY

FULTON, JOHN F.: Physiology of the nervous system. 667 pp., 140 figs. 1949, 3rd ed. Oxford University Press, New York.

ANATOMY

MAXIMOV, ALEXANDER A., and WILLIAM BLOOM: Textbook of histology. 616 pp., 580 figs. 1952, 6th ed. W. B. Saunders Co., Philadelphia.

SCHAEFFER, J. PARSONS: Morris' Human Anatomy. 1718 pp., 1220 figs. 1953, 11th ed. McGraw-Hill Book Co., 330 W. 42nd St., New York.

SOBOTTA, JOHANNES, and EDUARD UHLENHUTH: Atlas of human anatomy. 3 v., numerous color plates. 1955. Stechert-Hafner, New York.

CLINICAL

BING, ROBERT: Compendium of regional diagnosis, in lesions of the brain and spinal cord. 292 pp., 125 figs., 7 plates. 1940, 11th ed. C. V. Mosby Co., St. Louis.

BROCK, SAMUEL: Basis of clinical neurology. 393 pp. 1939, 2nd ed. Williams and Wilkins Co., Baltimore.

De JONG, RUSSELL H.: The neurologic examination. 1079 pp. 1950. P. B. Hoeber, New York.

GRINKER, ROY R., and PAUL C. BUCY: Neurology. 1138 pp., 393 figs. 1949, 4th ed. Charles C Thomas, Springfield, Ill.

MERRITT, H. HOUSTON; FRED A. METTLER and TRACY J. PUTNAM: Funda-mentals of clinical neurology. 289 pp., 96 figs. 1947. McGraw-Hill Book Co., 330 W. 42nd St., New York.

INDEX

The numbers in italics correspond to the numbers in the diachrome reconstruction. Special listings are given under: area, reflex, thalamic nuclei.

A

Abducent nucleus and nerve, 43, *88*
Aberrant pyramidal tracts (*see* corticobulbar tracts)
Abstracting, 112
Accessory (spinal accessory) nerve, 47, *264*
Accommodation reflex, 97
Acoustic nerve (VIII), 59, 62
  stria, 64
Adiadochokinesis, 139
Adrenalin, effect, 146
Adrenal medulla, 146
Adversive movements, 111, *8*
Alternating hemiplegia, 109
Alveus (*see* fornix)
Ambiguus nucleus, 46, *91*
Ammon's horn (*see* hippocampus)
Ampulla of semicircular canal, 52
Amygdala, 118, *228*
  salamander, 30
Analysis and synthesis of sensation, 92
  of visual sensation, 97
Angular acceleration, perception, 58
  gyrus, 102, *39*
Ansa lenticularis, 125, *135*
Anterior cerebral artery, 6
  chamber of eye, 67
  commissure, 103, *68*, *277*
    salamander, 30
  limb of internal capsule, 88, *218*
  lobe of cerebellum, 131, *82*
  nucleus of hypothalamus, 147
    of thalamus, 77, *111*
    projection, 117
  quadrangular lobule, 131
  thalamic radiation, 113, 117, *152*
Aortic arch reflex, 51
Aphasia, 101ff
  global, 102
  motor, 112
  nominal, 102
  semantic, 101
  syntactic, 102
Arachnoid, 4
Archicerebellum, 128

Archipallium (*see* olfactory cortex)
Archi- structures, 36
Arcuate associational fibers, 87
  fasciculus, 104, 112, 113, *271*
  fibers of cerebellum, 136
  nuclei of bulb, 134
  nucleus, 74, 77, 92, *149*
    medial division, 52, 116
Arcuatus nucleus of hypothalamus, 147
Area *1*, 93
  2, 94
  3, 92
  4, 106ff
  5, 96, *189*
  6, 109, *188*
  6, subdivisions, 111
  7, 96, *189*
  8, 111
  9, 113, 114, *187*
  10, 113, *187*
  11, *12*, 114, 148
  13-16, 115
  17, 71, 97
  18, 98
  19, 98
  20, *21*, 104
  22, 102
  23, *24*, 116
  28, 80, 117
  29, 116, 117
  31, *32*, 116
  35, *36*, 80, 105, 116
  37, 104
  38, 105
  39, 102
  40, 101
  41, 65, 100
  42, 100
  43, 96
  44, *45*, 112
  46, 113, 114
  51, 80, 117
Area, cortical, definition, 38
Areas of cerebral cortex, samples, 90, 91
Arnold's thalamic fasciculus (*see* thalamo-temporal fasciculus)

[175]

Lightning Source UK Ltd.
Milton Keynes UK
UKHW022112160223
417167UK00006B/109